My Book

This book belongs to

Name: _____

Copy right © 2019 MATH-KNOTS LLC

All rights reserved, no part of this publication may be reproduced, stored in any system or transmitted in any form, or by any means, electronic, mechanical, photocopying, recording, or otherwise without the written permission of MATH-KNOTS LLC.

Cover Design by :
Gowri Vemuri

First Edition :
June, 2020

Author :
Gowri Vemuri

Editor :
Ritvik Pothapragada

Questions: mathknots.help@gmail.com

Dedication

This book is dedicated to:

My Mom, who is my best critic, guide and supporter.

To what I am today, and what I am going to become tomorrow,

is all because of your blessings, unconditional affection and support.

This book is dedicated to the

strongest women of my life,

my dearest mom

and

to all those moms in this universe.

G.V.

INDEX

Notes	9 - 20
Literal equations	21 - 31
Absolute value equations	32 - 48
Absolute value inequalities	49 - 56
Find slope from the graph	57 - 61
Find the slope from the given points	62 - 64
Find the slope from the equation of straight line	65 - 67
Find the slope of parallel lines	68 - 70
Find the slope of perpendicular lines	71 - 73
Find the missing coordinate	74 - 76
Find the equation of the straight line	77 - 111
Graphing linear inequality	112 - 137

INDEX

Graphing absolute value equations	138 - 149
Graphing system of equations	150 - 172
Graphing system of inequalities	173 - 180
Answer Key	181 - 264

Advanced Algebra 1 notes

Expressions	Numerical expression	Algebraic/Variable expression
Expressions will not include equal sign	12^2 $8 + 10$ $4^2 - 16 + 9$	$6x + 17$ $5a^2 - 2b$
Equation Must include an equal sign; One side is equal to the other side	**Numerical equation** $9 + 8 = 17$ $5^2 - 11 + 2 = 16$	**Algebraic/variable equation** $6x + 5 = 35$ (x=__?__) $9x^2 = 225$ (x=__?__)

Solving Equations :

To solve an equation first add the like terms (if any) and then isolate the variable on to one side of the equation. To solve one - step equations, do the inverse operation to find the value of the variable.

Remember : Always perform the same operation on both sides of the equation to maintain the balance.
Inverse operation for addition is subtraction and vice versa.
Inverse operation for multiplication is division and vice versa.

Polynomial	Highest degree	1st Name by degree	2nd Name by number of terms
8	0	Constant	monomial
$9x + 8$	1	Linear	binomial
$-2x^2 + 7x - 11$	2	Quadratic	trinomial
$8x^3$	3	Cubic	monomial
$2x^4 + x^3$	4	Quartic	binomial
$3x^5 + 5x^3 + 12$	5	Quintic	trinomial
$2x^6 + x^5 - 3x^4 - 21$	6	6th degree	Polynomial with 4 terms
$5x^7 - x^5 + 4x^4 - 7x^2 + x - 3$	7	7th degree	Polynomial with 6 terms

ADVANCED ALGEBRA 1

Notes

A Polynomial in standard form is always written with terms in sequential order according to their highest degree of exponents (higher exponents to lower exponents).

Parts of the exponent:

This is read as "Seven to the fourth power"

Like Terms :

Two or more terms are said to be alike if they have the same variable and the same degree. Coefficients of like terms are not necessarily be same.

An expression is in its simplest form when

1. All like terms are combined.
2. All parentheses are opened and simplified.

Like Terms can combined by adding or subtracting their coefficients (pay attention to the positive and negative signs of the coefficient and apply rules of adding integers)

Combining like terms on the opposite side of the equal sign :

When the like terms are on opposite sides, we have to combine like terms by using the inverse operation and by undoing the equation.

Solving equations using the distributive property :

The number in front of the parentheses needs to be multiplied with every term within the parentheses. After the distribution and opening up the parentheses, combine like terms and solve.

Distributing with the negative sign :

Remember to apply the integer rules of positive and negative numbers while distributing.

```
+ X + = +
- X - = +
- X + = -
+ X - = -
```

ADVANCED ALGEBRA 1

Notes

Example 1:

$2x + 3 = x + 7$

$$\begin{array}{r} 2x + 3 = x + 7 \\ -x\ -3\ \ -x\ -3 \\ \hline x + 0 = 0 + 4 \end{array}$$

$x = 4$

⟵ Inverse operation for addition is subtraction

Example 2:

$7x + 5 = -3x + 25$

$$\begin{array}{r} 7x + 5 = -3x + 25 \\ 3x\ \ -5\ \ \ 3x\ \ \ -5 \\ \hline 10x + 0 = 0 + 20 \end{array}$$

$10x = 20$

$\dfrac{\cancel{10}x}{\cancel{10}} = \dfrac{\cancel{20}^{\,2}}{\cancel{10}}$

$\boxed{x = 2}$

⟵ Inverse operation for addition is subtraction and vice versa

⟵ Inverse operation for multiplication is division

Example 3:

$\dfrac{2x}{5} + 5 = 15$

$$\begin{array}{r} \dfrac{2x}{5} + 5 = 15 \\ -5\ \ \ -5 \\ \hline \dfrac{2x}{5} + 0 = 10 \end{array}$$

$\dfrac{2x}{5} = 10$

$\cancel{5} \cdot \dfrac{2x}{\cancel{5}} = 5 \cdot 10$

$\dfrac{\cancel{2}x}{\cancel{2}} = \dfrac{\cancel{50}^{\,25}}{\cancel{2}}$

$\boxed{x = 25}$

⟵ Inverse operation for addition is subtraction and vice versa

⟵ Inverse operation for division is multiplication

⟵ Inverse operation for multiplication is division

Inequality:

An inequality is a relation between two expressions that are not equal. As a mathematical statement an inequality states one side of the equation is less than, less than or equal to or greater than or greater than equal to the other side.

If the inequality has **less than** or **greater than** symbol,
1. The graph starts with the open circle.
2. For less than the graphing line goes toward the left.
3. For greater than the graphing line goes toward the right.

If the inequality has **less than or equal to** or **greater than or equal** to symbol,
1. The graph starts with the closed circle.
2. For less than or equal to the graphing line goes toward the left.
3. For greater than or equal to the graphing line goes toward the right.

Inequality statement	Inequality verbal expression	Inequality graph
x > -3 or -3 < x	x is greater than -3	
x < 3 or 3 > x	x is less than 3	
x >= -1 or -1 <= x	x is greater than or equal to -1	
x <= 1 or 1 <= x	x is less than or equal to 1	

Basic inequalities:

Solving inequalities is same as solving for an equation except for one special rule.

Compound Inequalities :

x < 0 or x ≥ 5 means all values less than 0 or 5 and more. In other words we are excluding the values 0,1,2,3,4.

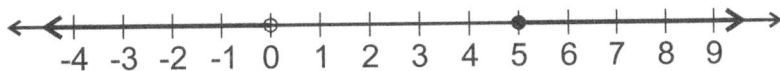

Absolute Value :

Absolute value of a number is its distance from 0. Since the distance cannot be negative absolute value is always positive.

$|7| = 7$ $|-11| = 11$

$|2.3| = 2.3$ $|-0.75| = 0.75$

Equations involving absolute values can be solved similar to regular algebraic equations solving. Absolute value should be treated as parentheses when applying PEDMAS rules.

Steps to solve absolute value equations.

Step 1 : Solve the expression within the absolute value.
(As applicable with PEDMAS rules)

Step 2 : Isolate the absolute value to one side of the equation.

Step 3 : Verify the value on the other side of the equation.
If the value is positive move to step 4.
If the value is negative there is no solution.

Step 4 : The expression inside the absolute value equals to positive and negative values of the other side of the equation.

Step 5 : Make the expression equal to positive value and solve for the variable.

Step 6 : Make the expression equal to negative value and solve for the variable.

Step 7 : The value obtained in step 5 and step 6 or the solutions to the absolute value equation.

Note : Absolute value equations can have two solutions. Since the absolute value of a number and its opposite are the same.
Absolute value can never be negative.

Absolute value Inequalities :

Absolute value inequalities are similar to absolute value equations.
Absolute value inequalities can have the below solutions
1. Two solutions
2. No solution
3. All real numbers

Steps to solve absolute value Inequalities are similar to solving the absolute value equations.

$|x| < a$ can be rewritten as $-a < x < a$ (where a is positive)
can also be written as $x < a$ **and** $x > -a$

$|x| \leq a$ can be rewritten as $-a \leq x \leq a$ (where a is positive)
can also be written as $x \leq a$ **and** $x \geq -a$

$|x| > a$ can be rewritten as $x > a$ **or** $x < -a$ (where a is positive)

Note : < or $\leq$ are represented by the word **and**
       \> or $\geq$ are represented by the word **or**

SLOPE :

Slope of a straight line is how far the line is away from the horizontal line in other words how slanted or angular the straight line is.

Slope of a straight line is the rate of change for a given set of points.
Example : (x_1, y_1) , (x_2, y_2)

$$\text{Slope } (m) = \frac{\text{Rise}}{\text{Run}} = \frac{\text{difference in the y coordinates}}{\text{difference in the x coordinates}} = \frac{y_2 - y_1}{x_2 - x_1} = \frac{y_1 - y_2}{x_1 - x_2}$$

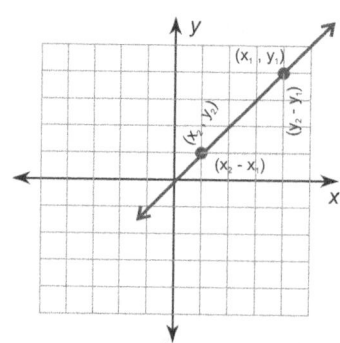

ADVANCED ALGEBRA 1

Notes

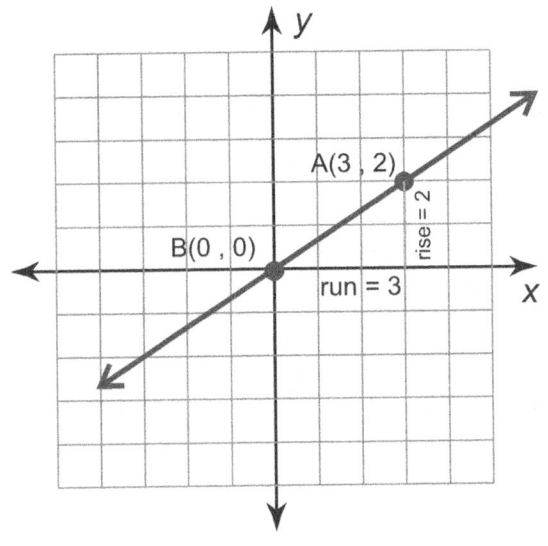

A straight line with a **positive** slope always **rises** from left to right

$A(3, 2)$ $B(0, 0)$
(x_2, y_2) (x_1, y_1)

$$\text{Slope} = \frac{y_2 - y_1}{x_2 - x_1} = \frac{2 - 0}{3 - 0} = \frac{2}{3}$$

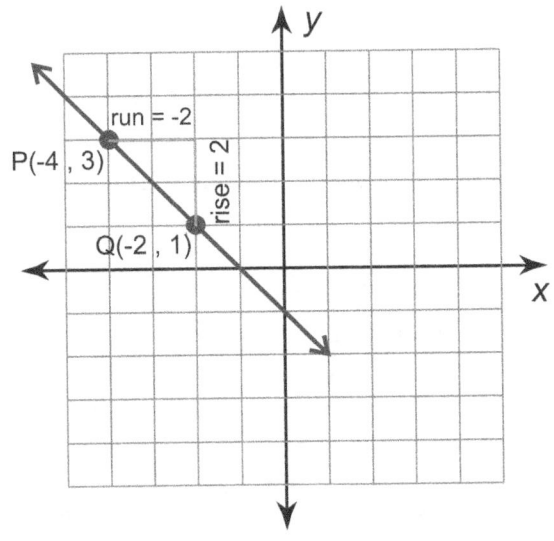

A straight line with a **negative** slope always **falls** from left to right

$P(-4, 3)$ $Q(-2, 1)$
(x_2, y_2) (x_1, y_1)

$$\text{Slope} = \frac{y_2 - y_1}{x_2 - x_1} = \frac{3 - 1}{-4 - (-2)} = \frac{2}{-4 + 2}$$

$$= \frac{2}{-2} = -1$$

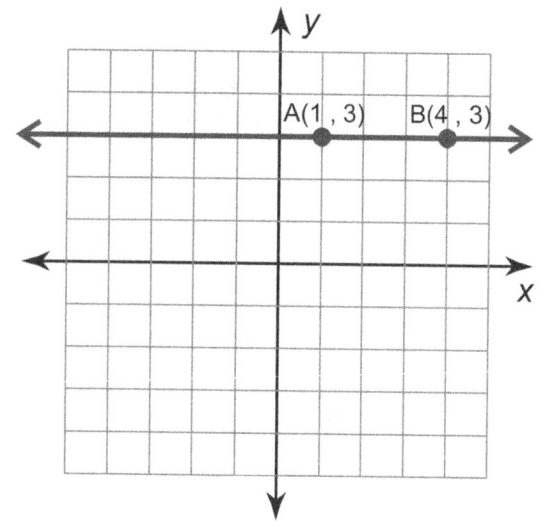

A straight line parallel to x axis always has a slope = 0

A(1, 3)   B(4, 3)
(x₂, y₂) (x₁, y₁)

Slope = $\dfrac{y_2 - y_1}{x_2 - x_1}$ = $\dfrac{3 - 3}{1 - 4}$ = $\dfrac{0}{-3}$ = 0

Note : A straight line of the form y = k, where k is a constant always has a zero slope.

Tip : When x coordinates are different and y coordinates are same slope is always zero.

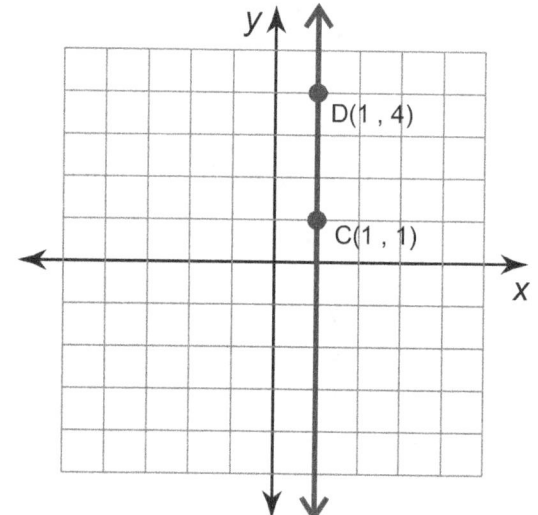

A straight line parallel to y axis always has an undefined slope

C(1, 1)   D(1, 4)
(x₂, y₂) (x₁, y₁)

Slope = $\dfrac{y_2 - y_1}{x_2 - x_1}$ = $\dfrac{1 - 4}{1 - 1}$ = $\dfrac{-3}{0}$ = undefined

Any number when you divide by 0 the value is undefined.

Note : A straight line of the form x = k, where k is a constant always has an undefined slope.

Tip : When x coordinates are same slope is undefined.

ADVANCED ALGEBRA 1

Slope intercept form of the straight line :

$y = mx + b$ is the slope intercept form of the straight line where m is the slope of the straight line and b is the y intercept.

Examples : $y = 7x - 1$; Slope = 7 , y intercept = -1

$y = -11x + 8$; Slope = -11 , y intercept = 8

Standard form of the straight line :

$ax + by = c$ is the Standard form of the straight line.

Example : $2x + 3y = 10$

Point slope form of the straight line :

$y - y_1 = m(x - x_1)$ is the point slope form of the straight line where m is the slope and (x_1, y_1) is any given point on the straight line.

Example : $y - 2 = 3(x - 5)$ then slope of the straight line is 3 and (5 , 2) is a given point on the straight line.

Example : Given two points A(2 , 3) and B(5 , 8). Find the equation of a straight line in point slope form. Also express it in slope intercept form and standard form.

Step 1 : Find the slope

$$m = \frac{y_2 - y_1}{x_2 - x_1} = \frac{8 - 3}{5 - 2} = \frac{5}{3}$$

Step 2 : Substitute the slope value obtained in step 1 and any one point A or B in the equation $y - y_1 = m(x - x_1)$

$$\boxed{y - 3 = \frac{5}{3}(x - 2)}$$ Equation of the straight line in point slope form

Step 3 : Simplify the equation obtained in step 2 to rewrite in the form of $y = mx + b$

$3(y - 3) = \frac{5}{\cancel{3}} \cancel{3} \cdot (x - 2)$

$3y - 9 = 5x - 10$

$3y - 9 + 9 = 5x - 10 + 9$

$\dfrac{3y}{3} = \dfrac{5x - 1}{3}$

$\boxed{y = \dfrac{5}{3}x - \dfrac{1}{3}}$ Equation of the straight line in slope intercept form

Step 4 : Ax + By = C is the standard form of the straight line

Lets use the point slope of the straight line obtained in step 2.

$y - 3 = \dfrac{5}{3}(x - 2)$

$3(y - 3) = \dfrac{5}{\cancel{3}}\cancel{3} \cdot (x - 2)$

$3y - 9 = 5x - 10$

$3y - 9 + 10 = 5x - 10 + 10$

$3y + 1 = 5x$

$3y + 1 - 3y = 5x - 3y$

$\boxed{5x - 3y = 1}$ Standard form of the straight line.

Important :
To find the x intercept of a given straight line substitute y = 0 and solve the equation. The value obtained is the x intercept.

Important :
To find the y intercept of a given straight line substitute x = 0 and solve the equation. The value obtained is the y intercept.

ADVANCED ALGEBRA 1

Notes

Parallel lines :

Two straight lines are parallel if and only if they are of the same slope but the y intercepts are different.

Example : y = 2x + 3 and y = 2x - 7 have the same slope = 3

Perpendicular lines :

Two straight lines are perpendicular if and only if their slopes are negative reciprocals of each other. Their y intercepts can be same or different.

Example : y = 5x + 9 and y = $-\frac{1}{5}$ x + 2 are perpendicular to each other as their slopes 5 and $-\frac{1}{5}$ are negative reciprocals of each other.

Important :
When the product of slopes of two straight lines equals to -1 then the lines are perpendicular to each other

---

Graphing the linear inequalities :

Graphing a linear inequality is similar to graphing any linear equation except ,

    1. If the inequality has ≤ or ≥ use a solid line to graph.

    2. If the inequality has < or > then use a dotted line to graph.

After plotting the solid or dotted line

    1. Test a point outside of the line to check if it is a solution of the given inequality.

    2. If the point tested is a solution then shade that side of the graph otherwise shade the opposite side.

    3. The shaded region and the solid line represents all the possible solutions of the given linear inequality. If the graph has the dotted line then only the shaded region represents all the solutions satisfying the linear inequality.

System of Linear equations :

Two or more linear equations can be solved by using elimination or substitution methods to find the common point or the points that satisfy both the equations.
The common points obtained are the solutions for the system of equations given.

Graphing System of Linear equations :

Two or more linear equations are graphed in the same way as graphing a linear equation.
The point at which the two lines intersect is the solution of the given pair of equations.
The solution can be cross verified by solving the equations as explained above.

Graphing System of Linear Inequalities :

Graphing two or more linear inequalities is similar to graphing a linear inequality and extending to more. The overlapping area of the inequalities graphed is the set of solutions that work for both system of linear inequalities graphed.

Graphing Absolute value of equations :

Graphing an absolute value of equation is similar to graphing an equation for y = absolute value of x consider all the possible order pair solutions to graph.
The point at which the graph turns to another direction is called the **vertex**.
The vertical line of symmetry for any graph is at its vertex.

ADVANCED ALGEBRA 1

Volume 1

Rewrite the below equations for the indicated variable

(1) $u = ka + ba$, for a

(2) $ukx = x + y$, for x

(3) $g = ca + ba$, for a

(4) $g + ca = ba$, for a

(5) $g = \dfrac{a+b}{ca}$, for a

(6) $u = kx + yx$, for x

(7) $z = \dfrac{a+b}{ma}$, for a

(8) $u = ka + ba$, for a

(9) $za = \dfrac{a+b}{m}$, for a

(10) $g + cx = yx$, for x

ADVANCED ALGEBRA 1

Volume 1

Rewrite the below equations for the indicated variable

(11) $ux = \dfrac{x+y}{k}$, for x

(12) $gc = \dfrac{x+y}{x}$, for x

(13) $gc = \dfrac{a+b}{a}$, for a

(14) $g = cx + yx$, for x

(15) $g = cx + yx$, for x

(16) $g = \dfrac{a+b}{ca}$, for a

(17) $u = \dfrac{x+y}{kx}$, for x

(18) $u + ka = ba$, for a

(19) $u + kx = yx$, for x

(20) $g = \dfrac{x+y}{cx}$, for x

Rewrite the below equations for the indicated variable

(21) $ukx = x + y$, for x

(22) $gca = a + b$, for a

(23) $zx = \dfrac{x+y}{m}$, for x

(24) $z = \dfrac{x+y}{mx}$, for x

(25) $z = ma + ba$, for a

(26) $ukx = x + y$, for x

(27) $zma = a + b$, for a

(28) $z + ma = ba$, for a

(29) $z + mx = yx$, for x

(30) $g + cx = yx$, for x

ADVANCED ALGEBRA 1

Rewrite the below equations for the indicated variable

(31) $z = \dfrac{xm + p}{xn}$, for x

(32) $x - c = dr + yx$, for x

(33) $xc = dr + yx$, for x

(34) $k - a = ba - vw$, for a

(35) $m - x = n + p + yx$, for x

(36) $x - m = n - p + yx$, for x

(37) $ac = dr + ba$, for a

(38) $z = xm + xn + p$, for x

(39) $a + m = ba - np$, for a

(40) $a + m = n + p + ba$, for a

Rewrite the below equations for the indicated variable

(41) $x + m = pn - yx$, for x

(42) $x - k = w + v + yx$, for x

(43) $am = p + n + ba$, for a

(44) $c - a = d + r + ba$, for a

(45) $x + c = d + r + yx$, for x

(46) $g = \dfrac{ac + r}{ad}$, for a

(47) $a + k = vw - ba$, for a

(48) $a - m = p + n + ba$, for a

(49) $kx = w - v + yx$, for x

(50) $xk = vw + yx$, for x

ADVANCED ALGEBRA 1

Volume 1

Rewrite the below equations for the indicated variable

(51) $z = am + an + p$, for a

(52) $a + c = d + r + ba$, for a

(53) $a + c = d - r + ba$, for a

(54) $g = \dfrac{xc + r}{xd}$, for x

(55) $g = ac + ad + r$, for a

(56) $xc = r - d + yx$, for x

(57) $ma = np + ba$, for a

(58) $u = xk + xw + xv$, for x

(59) $z = xm + xn + xp$, for x

(60) $a - m = p - n + ba$, for a

Rewrite the below equations for the indicated variable

(61) $u = \dfrac{a}{k}$, for a

(62) $u = \dfrac{k}{a}$, for a

(63) $z = mx$, for x

(64) $u = k - x$, for x

(65) $z = x + m$, for x

(66) $g = \dfrac{a}{c}$, for a

(67) $g = a - c$, for a

(68) $z = a - m$, for a

(69) $z = \dfrac{m}{x}$, for x

(70) $z = \dfrac{a}{m}$, for a

Rewrite the below equations for the indicated variable

(71) $g = \dfrac{c}{a}$, for a

(72) $g = c - x$, for x

(73) $u = a - k$, for a

(74) $z = \dfrac{x}{m}$, for x

(75) $g = cx$, for x

(76) $u = xk$, for x

(77) $z = m - a$, for a

(78) $u = \dfrac{x}{k}$, for x

(79) $z = am$, for a

(80) $z = x - m$, for x

Rewrite the below equations for the indicated variable

(81) $z = m - x$, for x

(82) $g = x - c$, for x

(83) $u = x + k$, for x

(84) $g = c + a$, for a

(85) $u = a + k$, for a

(86) $u = x - k$, for x

(87) $u = k - a$, for a

(88) $g = c - a$, for a

(89) $u = ak$, for a

(90) $g = \dfrac{c}{x}$, for x

ADVANCED ALGEBRA 1

Solve the below equations.

(91) $|p| = -5$

(92) $|p| = 9$

(93) $|n| = -9$

(94) $|a| = -12$

(95) $|n| = -2$

(96) $|b| = -10$

(97) $|a| = -15$

(98) $|n| = 6$

(99) $|x| = 7$

(100) $|x| = -11$

Solve the below equations.

(101) $|n| = -16$

(102) $|n| = -1$

(103) $|p| = 8$

(104) $|x| = 1$

(105) $|x| = 14$

(106) $|n| = 3$

(107) $|x| = 4$

(108) $|n| = 2$

(109) $|x| = -13$

(110) $|n| = -14$

ADVANCED ALGEBRA 1

Solve the below equations.

(111) $|x| = 11$

(112) $|v| = 13$

(113) $|v| = -3$

(114) $|k| = -6$

(115) $|a| = 5$

(116) $|n| = -4$

(117) $|n| = 12$

(118) $|x| = -8$

119) $|x| = 10$

120) $|x| = 0$

ADVANCED ALGEBRA 1

Volume 1

Solve the below equations.

(121) $|3v| = 9$

(122) $|x + 1| = 9$

(123) $\left|\dfrac{x}{8}\right| = 3$

(124) $\left|\dfrac{v}{9}\right| = 4$

(125) $|-9x| = 72$

(126) $|n + 10| = 16$

(127) $|-1 + x| = 3$

(128) $|5 + a| = 5$

(129) $|-3 + a| = 12$

(130) $|-6n| = 54$

ADVANCED ALGEBRA 1

Solve the below equations.

(131) $|x + 8| = 1$

(132) $|-1 + x| = 7$

(133) $\left|\dfrac{k}{9}\right| = 3$

(134) $|a - 1| = 8$

(135) $|-7 + v| = 2$

(136) $|n - 8| = 13$

(137) $\left|\dfrac{x}{9}\right| = 5$

(138) $|x - 9| = 1$

(139) $|9a| = 90$

(140) $|n - 6| = 3$

ADVANCED ALGEBRA 1

Volume 1

Solve the below equations.

(141) $|4n| = 32$

(142) $|6 + m| = 2$

(143) $|v + 8| = 2$

(144) $|p + 6| = 10$

(145) $|-3x| = 12$

(146) $|-5k| = 40$

(147) $|v + 2| = 8$

(148) $\left|\dfrac{r}{6}\right| = 2$

(149) $|7x| = 14$

(150) $|-1 + m| = 8$

Solve the below equations.

(151) $|4v - 9| = 33$

(152) $|11n + 6| = 104$

(153) $|-7 + x| = 11$

(154) $|12 + n| = 5$

(155) $|-p + 1| = 2$

(156) $|3r + 8| = 25$

(157) $|11 + 2k| = 13$

(158) $|8v + 10| = 30$

(159) $|-11p - 4| = 114$

(160) $|-9x + 7| = 97$

Solve the below equations.

(161) $|12 + 2k| = 10$

(162) $|4k + 5| = 45$

(163) $|2v - 3| = 21$

(164) $|4n + 12| = 4$

(165) $|9x + 1| = 19$

(166) $|10 + 2m| = 14$

(167) $|-10 + 5n| = 45$

(168) $|7 - 8x| = 47$

(169) $|-7 + 7a| = 70$

(170) $|-10 + n| = 20$

ADVANCED ALGEBRA 1

Volume 1

Solve the below equations.

(171) $|7 - 5x| = 48$

(172) $|12m - 2| = 106$

(173) $|-10n + 11| = 31$

(174) $|8x + 11| = 35$

(175) $|11x + 4| = 95$

(176) $|8x - 4| = 20$

(177) $|9 - 8x| = 15$

(178) $|9k - 5| = 41$

(179) $|-2 - 5m| = 53$

(180) $|9a + 6| = 30$

Solve the below equations.

(181) $\dfrac{|19+a|}{6} = 4$

(182) $-6 + \left|\dfrac{m}{4}\right| = -6$

(183) $|12r| - 12 = 204$

(184) $\left|\dfrac{x}{20}\right| + 10 = 11$

(185) $\dfrac{|15+v|}{4} = 3$

(186) $18|p - 13| = 0$

(187) $|-19 + b| + 3 = 40$

(188) $-16 - |2r| = 0$

(189) $\dfrac{|-19x|}{20} = -8$

(190) $\left|\dfrac{x}{17}\right| - 14 = -13$

Solve the below equations.

(191) $12\left|\dfrac{r}{20}\right| = 12$

(192) $-3\left|\dfrac{b}{20}\right| = -3$

(193) $|5x| - 18 = 32$

(194) $\left|\dfrac{k}{9}\right| - 15 = -13$

(195) $3 + |a + 10| = 21$

(196) $\dfrac{|20n|}{17} = 7$

(197) $\dfrac{|10b|}{16} = -4$

(198) $\left|\dfrac{k}{3}\right| + 5 = 10$

(199) $\dfrac{|x + 20|}{11} = 7$

(200) $-19\left|\dfrac{v}{7}\right| = -38$

Solve the below equations.

(201) $\dfrac{|-5n|}{7} = -10$

(202) $-1 - \left|\dfrac{x}{4}\right| = 4$

(203) $|n+6| + 6 = -8$

(204) $|k-9| + 1 = 15$

(205) $\dfrac{|-6a|}{6} = 2$

(206) $9|b+20| = 108$

(207) $16 + |17k| = 67$

(208) $\dfrac{|-16x|}{10} = 1$

(209) $7 - |x-20| = -21$

(210) $-13|14x| = -364$

ADVANCED ALGEBRA 1

Volume 1

Solve the below equations.

(211) $1 - 4|-1 + 13x| = -307$

(212) $-13|1 - a| + 11 = -210$

(213) $8|8 - 2v| + 16 = 160$

(214) $12 - 2|7x + 17| = -260$

(215) $-14|11b - 1| - 6 = -20$

(216) $7|-13n + 17| + 19 = 138$

(217) $8|5x + 5| + 13 = 253$

(218) $15|16x + 20| + 2 = 302$

(219) $13|14k + 6| + 2 = 80$

(220) $3 - 8|16 + 14n| = -349$

Solve the below equations.

(221) $2 + 3|4k + 7| = 203$

(222) $3|6r + 17| - 6 = 297$

(223) $5 - 16|14 + 11n| = -219$

(224) $18|10n + 14| + 17 = 89$

(225) $20 - 6|5x + 5| = -70$

(226) $6 + 4|15x + 2| = 314$

(227) $16|-2b + 12| + 20 = 436$

(228) $5|9 - 3k| + 12 = 162$

(229) $4 - 8|4 - 15m| = -388$

(230) $-4|20b + 8| + 4 = -28$

Solve the below equations.

(231) $19 + 12|10 - 7x| = 403$

(232) $5|-13x + 18| - 3 = 232$

(233) $11|12x + 14| + 5 = 379$

(234) $2|12n + 16| - 2 = 302$

(235) $20 + 14|2x + 5| = 146$

(236) $4|19 - 15n| + 11 = 87$

(237) $18|-6 + 13b| - 6 = 336$

(238) $10|-10 + 2n| + 20 = 260$

(239) $13|19v - 16| + 5 = 291$

(240) $15|4 - 20n| + 9 = 69$

Solve the below equations.

(241) $-12|7x - 16| - 9 = -237$

(242) $2 - 2|20p + 2| = -434$

(243) $18|17x - 6| + 8 = 422$

(244) $1 + 6|-19 + 13x| = 43$

(245) $17 - 8|-15b - 7| = -399$

(246) $|-17 - 8n| - 8 = 79$

(247) $4 + 6|9r - 18| = 166$

(248) $6|16 - 11n| + 16 = 118$

(249) $20|15x - 4| + 16 = 96$

(250) $4 + 2|17n + 15| = 8$

Solve the below equations.

(251) $-6|5 + 6p| - 7 = -85$

(252) $-6 - 2|18r - 1| = -332$

(253) $-13|7v - 8| + 7 = -162$

(254) $4|12x + 1| + 19 = 255$

(255) $5 + 12|16n - 5| = 257$

(256) $19|-8v - 7| + 16 = 35$

(257) $12|-10 + 3n| - 6 = 126$

(258) $-7|7n - 4| - 12 = -33$

(259) $-16 - 12|-13n - 19| = -100$

(260) $15|-17n + 8| + 18 = 153$

Solve the below inequalities and graph its solution.

(261) $\left|\dfrac{n}{3}\right| + 5 \geq 6$

(262) $\left|\dfrac{k}{7}\right| - 4 \leq -3$

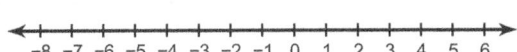

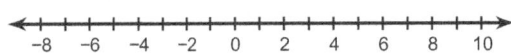

(263) $6 + |-4 + r| < 20$

(264) $9 + \left|\dfrac{a}{10}\right| \geq 10$

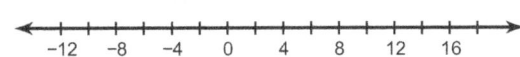

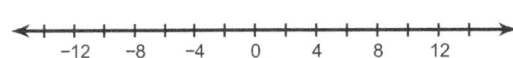

(265) $|v + 8| + 7 < 8$

(266) $7 + |b - 10| > 16$

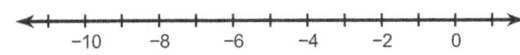

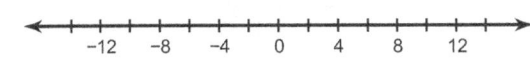

(267) $|3x| - 1 \geq 29$

(268) $4 + \left|\dfrac{n}{2}\right| > 7$

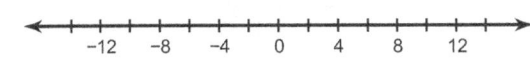

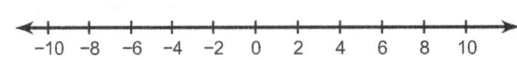

Solve the below inequalities and graph its solution.

(269) $|-3n| + 1 > 10$

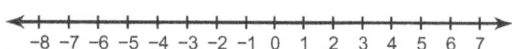

(270) $-8|a - 4| > -88$

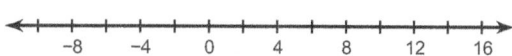

(271) $\left|\dfrac{k}{5}\right| - 9 \leq -7$

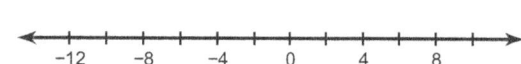

(272) $2|n + 8| > 26$

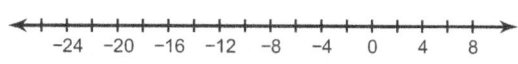

(273) $-8|m - 9| > -96$

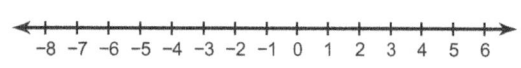

(274) $-6|a + 4| \leq -12$

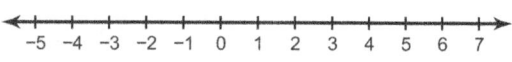

(275) $\left|\dfrac{a}{3}\right| + 3 > 4$

(276) $6|-4 + x| < 12$

Solve the below inequalities and graph its solution.

(277) $|v - 3| + 5 \leq 7$

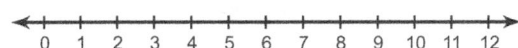

(278) $|r + 5| - 4 < 7$

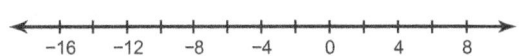

(279) $\dfrac{|m + 9|}{3} < 3$

(280) $-5 + \left|\dfrac{b}{4}\right| < -3$

(281) $|4 + b| - 6 < 1$

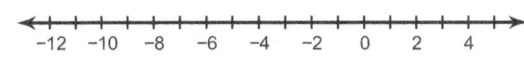

(282) $-2\left|\dfrac{b}{8}\right| \geq -2$

(283) $|-3b| + 2 \leq 32$

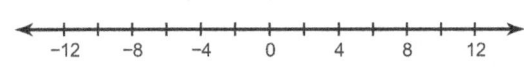

(284) $-4 + |a - 7| \geq 12$

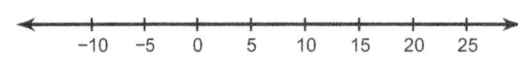

ADVANCED ALGEBRA 1

Volume 1

Solve the below inequalities and graph its solution.

(285) $|9 + a| + 7 > 19$

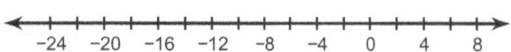

(286) $-10 + |-9 + n| < -6$

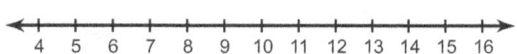

(287) $\dfrac{|k + 1|}{8} > 4$

(288) $|a - 10| - 6 \leq 13$

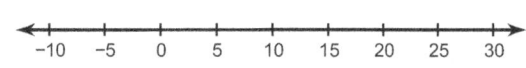

(289) $|-2r| + 7 > 15$

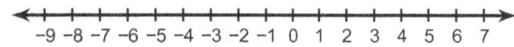

(290) $|2p| + 1 \geq 17$

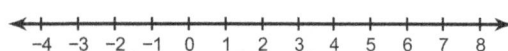

(291) $-2 - 10|9 - 6x| > -92$

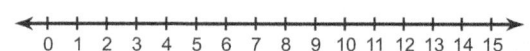

(292) $6|-7 + x| - 7 \leq 29$

 **ADVANCED ALGEBRA 1**

Volume 1

Solve the below inequalities and graph its solution.

(293) $9 + 9|10r + 7| > 72$

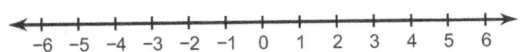

(294) $5|8x - 3| + 5 \geq 60$

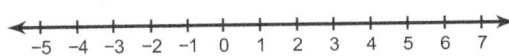

(295) $8 - 3|-x - 2| < -4$

(296) $9 + 8|7 - 9x| \leq 65$

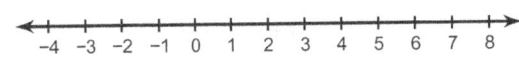

(297) $6|7 - 3x| - 6 > 78$

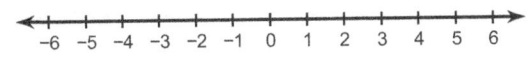

(298) $10|9 + 7v| - 1 > 89$

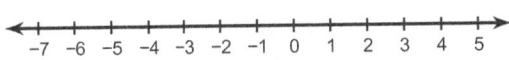

(299) $8 - 7|-6 - 6n| > -76$

(300) $3|8n + 7| + 1 < 118$

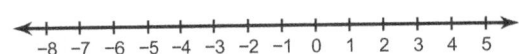

Solve the below inequalities and graph its solution.

(301) $10 - 4|8k + 6| > -62$

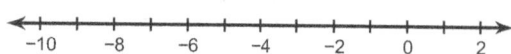

(302) $3 + 10|8x + 3| > 33$

(303) $3|m + 9| - 3 \geq 39$

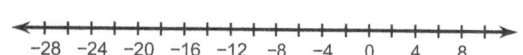

(304) $-4 + 2|4 + 5a| > 14$

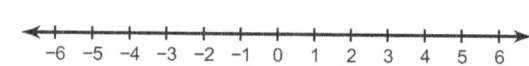

(305) $2 - 10|6m + 4| \leq -38$

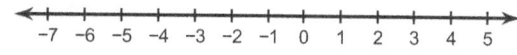

(306) $2|-8 - 9x| + 8 > 42$

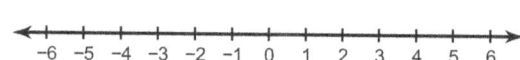

(307) $8|3n + 10| + 2 \geq 10$

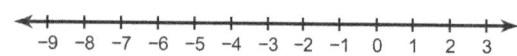

(308) $6 - 5|-6 + 4x| \leq -24$

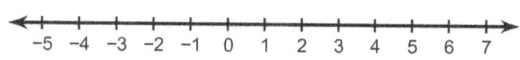

Solve the below inequalities and graph its solution.

(309) $-3|5n-8| - 8 \le -62$

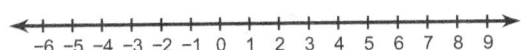

(310) $5|10 + 3p| + 4 < 74$

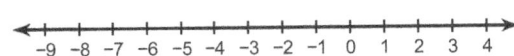

(311) $7 + 5|-10r - 4| < 87$

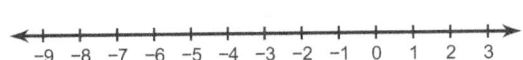

(312) $-7|9 - 3x| + 6 < -57$

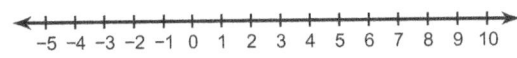

(313) $-9 - |8 - 10x| \ge -17$

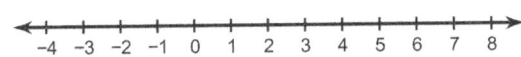

(314) $2 + 10|8k + 7| \le 72$

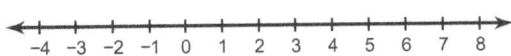

(315) $-7 + 8|10 + 7n| > 25$

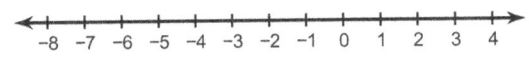

(316) $9 - 4|2 - 4x| \le -95$

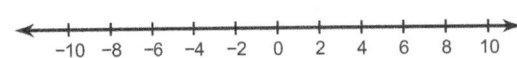

Solve the below inequalities and graph its solution.

(317) $5 + 7\left|-9k + 9\right| > 68$

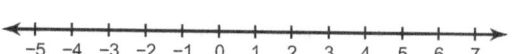

(318) $3\left|5r - 6\right| - 9 \leq 33$

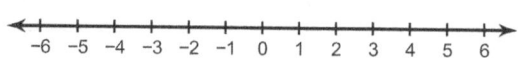

(319) $5\left|4 - 9n\right| - 8 > 57$

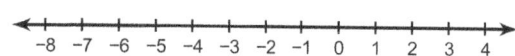

(320) $1 + 3\left|4x + 2\right| \leq 19$

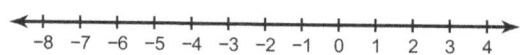

Find the slope of the straight line as shown in the graph below.

(321)
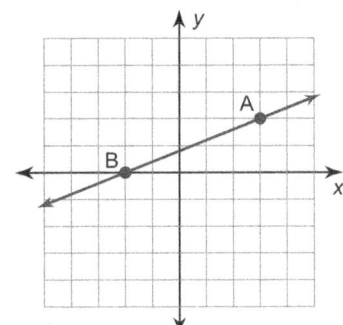

(322)
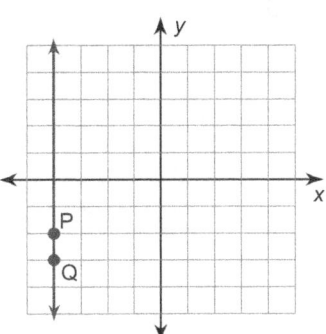

(323)
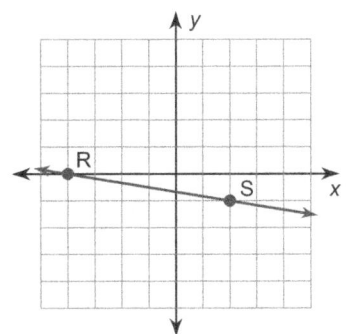

(324)
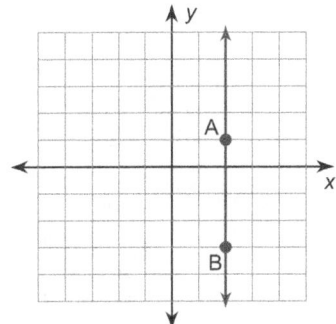

(325)
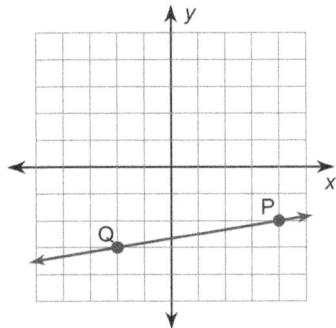

(326)
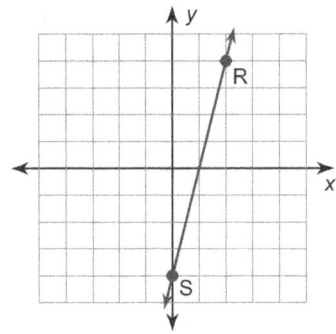

Find the slope of the straight line as shown in the graph below.

(327)
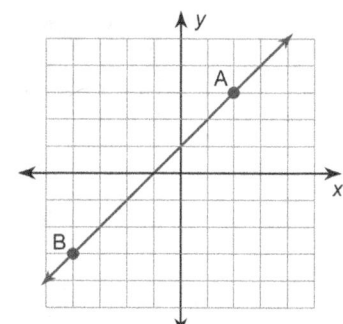

(328)
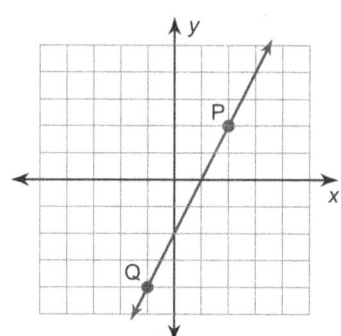

(329)
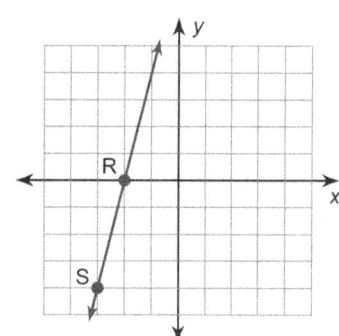

(330)
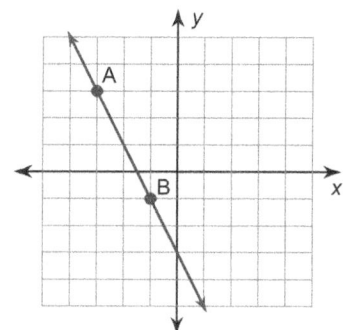

(331)
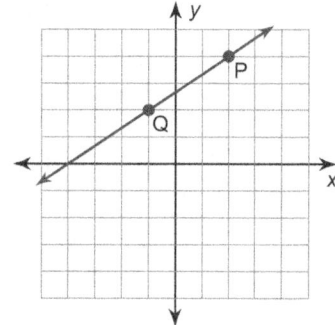

(332)
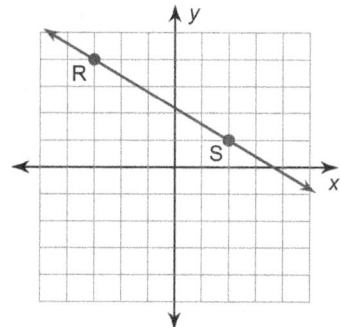

Find the slope of the straight line as shown in the graph below.

(333)
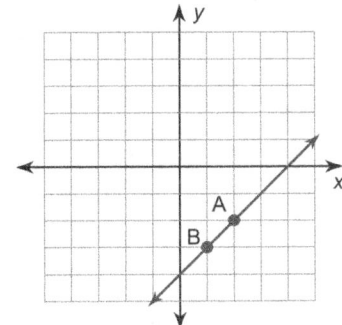

(334)
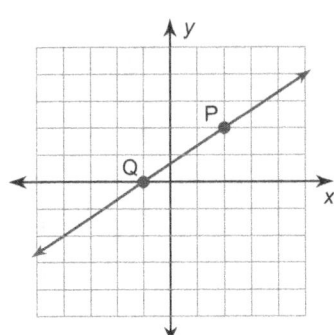

(335)
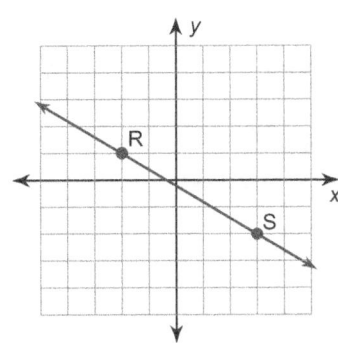

(336)
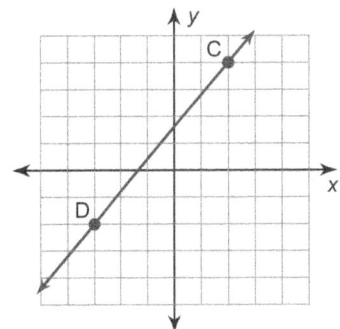

(337)
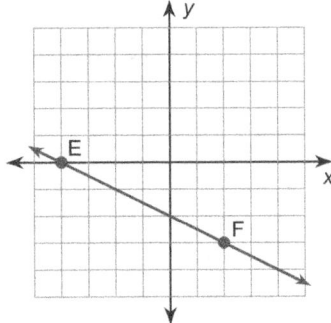

(338)
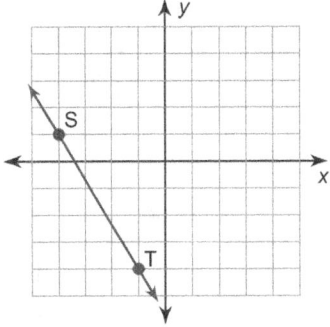

Find the slope of the straight line as shown in the graph below.

(339)
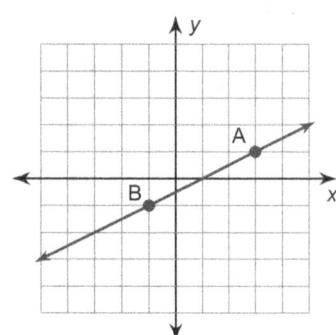

(340)
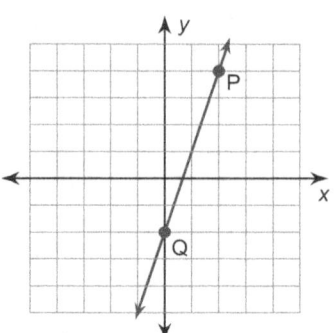

(341)
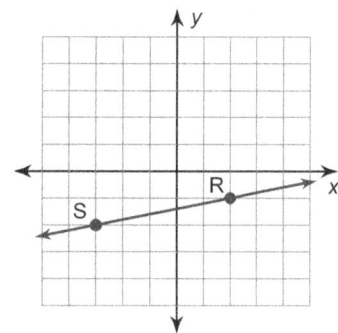

(342)
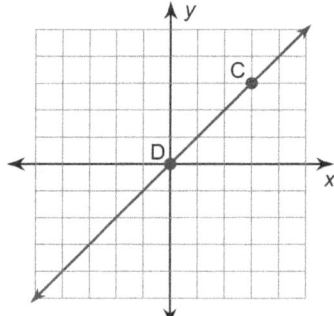

(343)
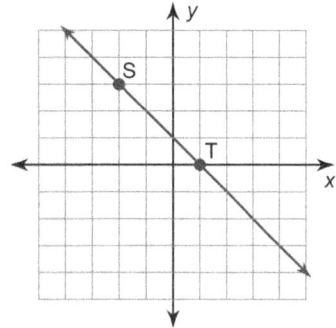

(344)
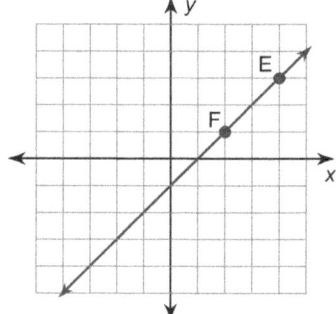

Find the slope of the straight line as shown in the graph below.

(345)

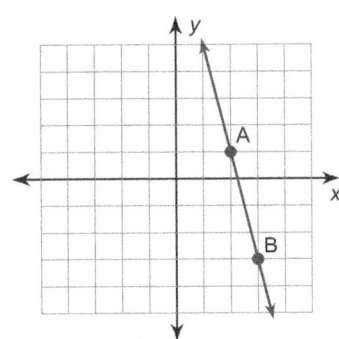

Find the slope of the straight line passing through the pair of points as given below.

(346) A(3, −9), B(−18, −15)

(347) A(0, −2), B(11, 4)

(348) A(17, −2), B(−1, 19)

(349) A(−1, −3), B(1, 18)

(350) A(−12, −15), B(9, −18)

(351) A(6, 17), B(−19, −11)

(352) A(−19, −18), B(9, −11)

(353) A(−6, 13), B(−1, 14)

(354) A(1, −12), B(−14, −15)

(355) A(6, −13), B(10, 8)

Find the slope of the straight line passing through the pair of points as given below.

(356) A(20, 19), B(17, −9)

(357) A(3, 15), B(−18, −1)

(358) A(6, 6), B(−17, 10)

(359) A(16, −13), B(−20, 15)

(360) A(13, 4), B(3, 11)

(361) A(4, 1), B(−5, 11)

(362) A(6, 6), B(−5, −14)

(363) A(1, 17), B(−9, −17)

(364) A(−8, −9), B(−3, −6)

(365) A(−9, 19), B(8, 2)

Find the slope of the straight line passing through the pair of points as given below.

(366)   A(-15, 8), B(3, 20)

(367)   A(8, 12), B(20, -8)

(368)   A(14, 2), B(-15, 8)

(369)   A(-6, -13), B(10, -1)

(370)   A(-9, 0), B(-14, -4)

Find the slope of each straight line as given below.

(371) $y = 3x - 5$

(372) $y = 2x - 4$

(373) $y = \dfrac{2}{3}x + 5$

(374) $y = -2x + 5$

(375) $y = 1$

(376) $y = x + 2$

(377) $y = -\dfrac{5}{3}x + 2$

(378) $x = -4$

(379) $y = 0$

(380) $x = 5$

Find the slope of each straight line as given below.

(381) $y = 7x + 5$

(382) $y = -\dfrac{5}{2}x + 1$

(383) $y = -\dfrac{4}{3}x - 1$

(384) $x = -2$

(385) $y = x - 4$

(386) $y = \dfrac{3}{2}x$

(387) $y = 4$

(388) $y = -\dfrac{2}{5}x - 2$

(389) $y = 10x - 5$

(390) $y = 3$

Find the slope of each straight line as given below.

(391) $y = 6x - 5$

(392) $y = -\dfrac{7}{3}x + 4$

(393) $y = -x$

(394) $y = \dfrac{1}{2}x + 1$

(395) $x = 4$

ADVANCED ALGEBRA 1

Find the slope of the parallel line to each straight line as given below.

(396) $-3 + 2x = -3y$

(397) $0 = -x + \dfrac{1}{5} + \dfrac{1}{5}y$

(398) $-y + x = 1$

(399) $10x + 24 = 8y$

(400) $0 = -5x + 9 - 3y$

(401) $-x = 2 + y$

(402) $25 + 3x - 5y = 0$

(403) $3 = -y$

(404) $\dfrac{16}{5}x = -2y + 10$

(405) $-3y = -3 - 4x$

Find the slope of the parallel line to each straight line as given below.

(406) $y + 1 = 3x$

(407) $2x + 4y = 4$

(408) $-3y - 9 + 2x = 0$

(409) $4x + 6 = -6y$

(410) $6x - 3y = 3$

(411) $-1 = -\dfrac{1}{4}y + \dfrac{1}{4}x$

(412) $3 - y = 2x$

(413) $45 + 9y = 24x$

(414) $-4 = -2x + y$

(415) $-y + x + 3 = 0$

Find the slope of the parallel line to each straight line as given below.

(416) $8 + x = -4y$

(417) $-y - 5x = 5$

(418) $3 = -5x - 3y$

(419) $\frac{2}{7} y = 2x + \frac{10}{7}$

(420) $3 + 2x = y$

(421) $2x + y = 1$

(422) $x - 4y = 4$

(423) $\frac{1}{4} y + \frac{3}{4} = -x$

(424) $9 + 2x = 3y$

(425) $0 = x - \frac{2}{9} y - \frac{10}{9}$

ADVANCED ALGEBRA 1

Volume 1

Find the slope of the perpendicular line to each straight line as given below.

(426) $x + 3 + \dfrac{3}{2} y = 0$

(427) $-2 + 2y = 4x$

(428) $0 = 2 + y - x$

(429) $0 = 18 + 3x + 9y$

(430) $10 = 5y - x$

(431) $-1 - \dfrac{1}{5} y - \dfrac{2}{5} x = 0$

(432) $4x + 5 = 5y$

(433) $0 = 2y + 10 + 5x$

(434) $4 - x = 0$

(435) $-1 - y = -5x$

Find the slope of the perpendicular line to each straight line as given below.

(436) $-4 = -6x + y$

(437) $-2 + y = 0$

(438) $2y - 2 = -x$

(439) $0 = -2y - 2$

(440) $-2x + 10 = -5y$

(441) $-y + \dfrac{3}{2} x = -2$

(442) $0 = -x - \dfrac{1}{5} y + \dfrac{1}{5}$

(443) $4 + \dfrac{1}{4} x = y$

(444) $-10 = x + 5y$

(445) $x - \dfrac{5}{8} y = -\dfrac{15}{8}$

Find the slope of the perpendicular line to each straight line as given below.

(446) $5y - 6x = 25$

(447) $18x = 15y + 75$

(448) $-\dfrac{5}{8} y = x - \dfrac{25}{8}$

(449) $3 + y = -2x$

(450) $0 = -6x - 2y - 4$

(451) $0 = 6x + 8y + 16$

(452) $-5 - x = 0$

(453) $x = -4$

(454) $-2x - 2y - 8 = 0$

(455) $-6y + 4x - 12 = 0$

Find the missing coordinate x or y for the below given the slope of the straight line

(456) $(31, y)$ and $(3, 35)$;

slope: 0

(457) $(x, 18)$ and $(7, 38)$;

slope: $-\dfrac{4}{5}$

(458) $(x, 29)$ and $(9, 32)$;

slope: $\dfrac{1}{3}$

(459) $(26, y)$ and $(33, -23)$;

slope: 1

(460) $(-31, -20)$ and $(-22, y)$;

slope: $-\dfrac{26}{9}$

(461) $(-36, y)$ and $(-41, 38)$;

slope: $-\dfrac{27}{5}$

(462) $(3, 21)$ and $(-15, y)$;

slope: $\dfrac{10}{9}$

(463) $(-1, 46)$ and $(20, y)$;

slope: $-\dfrac{23}{7}$

(464) $(x, -2)$ and $(-30, 43)$;

slope: $\dfrac{45}{4}$

(465) $(-42, y)$ and $(-36, -27)$;

slope: $-\dfrac{25}{3}$

ADVANCED ALGEBRA 1

Volume 1

Find the missing coordinate x or y for the below given the slope of the straight line

(466) $(31, y)$ and $(3, 35)$;

slope: $-\dfrac{18}{7}$

(467) $(-26, -36)$ and $(-22, y)$;

slope: 3

(468) $(-29, 22)$ and $(x, 38)$;

slope: $\dfrac{2}{3}$

(469) $(41, -31)$ and $(5, y)$;

slope: $-\dfrac{13}{9}$

(470) $(-7, y)$ and $(-5, -15)$;

slope: $-\dfrac{53}{2}$

(471) $(x, -47)$ and $(29, 13)$;

slope: $\dfrac{5}{2}$

(472) $(-5, y)$ and $(-15, 45)$;

slope: $-\dfrac{27}{10}$

(473) $(33, 46)$ and $(x, -43)$;

slope: $-\dfrac{89}{8}$

(474) $(x, 7)$ and $(-33, 24)$;

slope: undefined

(475) $(45, y)$ and $(46, -11)$;

slope: -29

ADVANCED ALGEBRA 1

Find the missing coordinate x or y for the below given the slope of the straight line

(476) $(x, -29)$ and $(47, 10)$;
slope: $\dfrac{3}{7}$

(477) $(19, y)$ and $(17, -39)$;
slope: 40

(478) $(x, 16)$ and $(23, -30)$;
slope: $\dfrac{46}{3}$

(479) $(39, y)$ and $(43, 6)$;
slope: $\dfrac{35}{4}$

(480) $(36, -38)$ and $(41, y)$;
slope: $-\dfrac{11}{5}$

(481) $(-31, y)$ and $(-13, 1)$;
slope: $\dfrac{4}{9}$

(482) $(x, 0)$ and $(14, 21)$;
slope: $\dfrac{7}{3}$

(483) $(x, -10)$ and $(33, -25)$;
slope: 15

(484) $(x, -17)$ and $(26, 29)$;
slope: $-\dfrac{46}{9}$

(485) $(3, y)$ and $(-21, -31)$;
slope: $\dfrac{13}{8}$

ADVANCED ALGEBRA 1

Volume 1

Write the standard form of the equation of the straight line given the slope and y-intercept.

(486) Slope = $-\dfrac{3}{5}$, y-intercept = 0

(487) Slope = $\dfrac{2}{3}$, y-intercept = −2

(488) Slope = $-\dfrac{1}{2}$, y-intercept = 2

(489) Slope = $\dfrac{1}{3}$, y-intercept = −3

(490) Slope = $-\dfrac{7}{3}$, y-intercept = 2

(491) Slope = $-\dfrac{2}{3}$, y-intercept = 5

(492) Slope = $\dfrac{5}{3}$, y-intercept = 4

(493) Slope = $-\dfrac{3}{2}$, y-intercept = 3

(494) Slope = −9, y-intercept = −4

(495) Slope = 5, y-intercept = 0

Write the standard form of the equation of the straight line given the slope and y-intercept.

(496) Slope = 2 , y-intercept = 3

(497) Slope = −3 , y-intercept = −3

(498) Slope = $\dfrac{2}{5}$, y-intercept = −2

(499) Slope = $\dfrac{9}{5}$, y-intercept = 5

(500) Slope = 0 , y-intercept = −5

(501) Slope = −4 , y-intercept = 2

(502) Slope = −1 , y-intercept = 2

(503) Slope = $\dfrac{1}{2}$, y-intercept = −2

(504) Slope = $\dfrac{5}{3}$, y-intercept = −2

(505) Slope = $-\dfrac{5}{2}$, y-intercept = −2

ADVANCED ALGEBRA 1

Write the standard form of the equation of the straight line given the slope and y-intercept.

(506) Slope = 2 , y-intercept = −1

(507) Slope = $-\dfrac{3}{5}$, y-intercept = −3

(508) Slope = $-\dfrac{6}{5}$, y-intercept = −2

(509) Slope = $\dfrac{3}{5}$, y-intercept = 4

(510) Slope = $-\dfrac{8}{5}$, y-intercept = −4

Write the slope intercept form of the equation of the straight line given the slope and y-intercept.

(511) Slope = $\frac{1}{2}$, y-intercept = −2

(512) Slope = −3, y-intercept = 0

(513) Slope = $\frac{1}{5}$, y-intercept = 0

(514) Slope = $-\frac{1}{5}$, y-intercept = −4

(515) Slope = 0, y-intercept = −2

(516) Slope = $\frac{3}{5}$, y-intercept = −1

(517) Slope = $-\frac{2}{5}$, y-intercept = 3

(518) Slope = −3, y-intercept = 4

(519) Slope = −8, y-intercept = 4

(520) Slope = $\frac{1}{5}$, y-intercept = −1

Write the slope intercept form of the equation of the straight line given the slope and y-intercept.

(521) Slope = 4 , y-intercept = −4

(522) Slope = $-\dfrac{1}{5}$, y-intercept = −5

(523) Slope = 1 , y-intercept = −2

(524) Slope = $\dfrac{1}{5}$, y-intercept = 2

(525) Slope = $\dfrac{8}{3}$, y-intercept = −5

(526) Slope = 1 , y-intercept = 4

(527) Slope = 2 , y-intercept = 3

(528) Slope = 7 , y-intercept = 5

(529) Slope = $-\dfrac{7}{4}$, y-intercept = 5

(530) Slope = −6 , y-intercept = 1

Write the slope intercept form of the equation of the straight line given the slope and y-intercept.

(531) Slope = $\dfrac{5}{3}$, y-intercept = -1

(532) Slope = $\dfrac{1}{5}$, y-intercept = -3

(533) Slope = $\dfrac{7}{4}$, y-intercept = -4

(534) Slope = $-\dfrac{1}{2}$, y-intercept = 0

(535) Slope = 4, y-intercept = -1

ADVANCED ALGEBRA 1

Volume 1

Write the standard form of the equation of the straight line passing the given point and having the slope.

(536) $A(-1, -4)$, slope = 5

(537) $A(4, -4)$, slope = $-\dfrac{3}{4}$

(538) $A(2, -5)$, slope = -3

(539) $A(-3, 2)$, slope = $-\dfrac{4}{3}$

(540) $A(-1, -5)$, slope = 5

(541) $A(3, -5)$, slope = $-\dfrac{7}{3}$

(542) $A(-5, 0)$, slope = $\dfrac{4}{5}$

(543) $A(3, 3)$, slope = undefined

(544) $A(1, 5)$, slope = 7

(545) $A(1, -3)$, slope = -6

ADVANCED ALGEBRA 1

Volume 1

Write the standard form of the equation of the straight line passing the given point and having the slope.

(546) $A(0, 2)$, slope = $\dfrac{3}{4}$

(547) $A(-3, 3)$, slope = $-\dfrac{7}{3}$

(548) $A(-2, -2)$, slope = $-\dfrac{1}{2}$

(549) $A(-2, -3)$, slope = $\dfrac{4}{3}$

(550) $A(-4, 3)$, slope = -1

(551) $A(2, 2)$, slope = $\dfrac{7}{2}$

(552) $A(-3, -4)$, slope = undefined

(553) $A(1, -1)$, slope = 4

(554) $A(-2, -5)$, slope = 5

(555) $A(-4, 1)$, slope = $\dfrac{1}{9}$

Write the standard form of the equation of the straight line passing the given point and having the slope.

(556) A(−3, 1), slope = 1

(557) A(−3, −2), slope = 3

(558) A(−3, 0), slope = −2

(559) A(3, 4), slope = $-\dfrac{1}{3}$

(560) A(4, 4), slope = $\dfrac{7}{2}$

ADVANCED ALGEBRA 1

Volume 1

Rewrite the below given equations into slope-intercept form of the straight line.

(561) $0 = 6x + 4 - y$

(562) $25 = 5y - 3x$

(563) $-5y = 10 - 2x$

(564) $3 + \frac{3}{2} y = x$

(565) $25 = 5y - 7x$

(566) $-y = x$

(567) $-3 + x = \frac{3}{4} y$

(568) $0 = 3y - 30 - \frac{15}{2} x$

(569) $10 = 4x - 2y$

(570) $4 + x = 0$

Rewrite the below given equations into slope-intercept form of the straight line.

(571) $-2x = -10$

(572) $\dfrac{9}{5}x = -y - 5$

(573) $-x - 2y = -4$

(574) $-2 - 5x = -y$

(575) $0 = -2x + 3y - 6$

(576) $0 = 45 - 21x + 15y$

(577) $-y - 4 + 3x = 0$

(578) $0 = x + 2$

(579) $0 = 5x - 5 + y$

(580) $-3x + y = 2$

Rewrite the below given equations into slope-intercept form of the straight line.

(581) $6x - 5y = 25$

(582) $-x - y = 4$

(583) $3 = -x$

(584) $-8y = 2x + 40$

(585) $6 = -2y - 3x$

ADVANCED ALGEBRA 1

Write the point-slope form of the equation of the line passing through the given point with the given slope.

(586) A(−2, −5), slope = undefined

(587) A(0, 5), slope = −9

(588) A(−4, −4), slope = $\dfrac{4}{3}$

(589) A(2, −5), slope = −5

(590) A(3, −2), slope = −2

(591) A(5, −4), slope = 0

(592) A(2, 4), slope = 3

(593) A(−2, −5), slope = 2

(594) A(2, −5), slope = undefined

(595) A(−3, 0), slope = $-\dfrac{5}{7}$

ADVANCED ALGEBRA 1

Volume 1

Write the point-slope form of the equation of the line passing through the given point with the given slope.

(596) $A(1, 4)$, slope = -1

(597) $A(-4, -5)$, slope = $\dfrac{5}{4}$

(598) $A(1, 5)$, slope = undefined

(599) $A(1, 4)$, slope = 3

(600) $A(-1, 5)$, slope = -9

(601) $A(-2, -1)$, slope = $-\dfrac{1}{2}$

(602) $A(2, 5)$, slope = $\dfrac{9}{2}$

(603) $A(-4, -4)$, slope = $\dfrac{5}{4}$

(604) $A(-4, 4)$, slope = -2

(605) $A(0, -2)$, slope = $-\dfrac{7}{2}$

Write the point-slope form of the equation of the line passing through the given point with the given slope.

(606) $A(0, 4)$, slope $= -\dfrac{4}{3}$

(607) $A(-3, -3)$, slope $= 2$

(608) $A(5, -2)$, slope $= \dfrac{2}{5}$

(609) $A(-4, -1)$, slope $= -\dfrac{1}{4}$

(610) $A(3, -2)$, slope $= -\dfrac{1}{3}$

ADVANCED ALGEBRA 1

Volume 1

Write the slope-intercept form of the equation of the line passing through the given points.

(611) A(−2, 0) and B(−1, 5)

(612) A(1, 0) and B(0, −5)

(613) A(2, 4) and B(5, 2)

(614) A(0, −3) and B(1, 0)

(615) A(5, 5) and B(0, 1)

(616) A(0, −5) and B(−2, −4)

(617) A(0, −4) and B(4, −2)

(618) A(1, 2) and B(−4, −4)

(619) A(0, −3) and B(3, 3)

(620) A(1, 4) and B(0, 5)

ADVANCED ALGEBRA 1

Volume 1

Write the slope-intercept form of the equation of the line passing through the given points.

(621) A(4, 2) and B(−3, −5)

(622) A(4, 0) and B(0, −3)

(623) A(−1, −4) and B(−3, 3)

(624) A(−2, −2) and B(−3, 2)

(625) A(−4, −3) and B(1, −1)

(626) A(3, 2) and B(−2, 4)

(627) A(−5, 3) and B(4, −3)

(628) A(5, 5) and B(−2, 2)

(629) A(3, 1) and B(0, 2)

(630) A(1, −2) and B(4, −3)

ADVANCED ALGEBRA 1

Volume 1

Write the slope-intercept form of the equation of the line passing through the given points.

(631) A(−2, 4) and B(4, −3)

(632) A(0, 4) and B(−4, 3)

(633) A(1, −1) and B(3, −5)

(634) A(0, 4) and B(3, 1)

(635) A(0, 4) and B(4, −3)

(636) A(4, 2) and B(−1, −1)

(637) A(4, 4) and B(0, 0)

(638) A(0, −5) and B(3, 2)

(639) A(0, −5) and B(−5, 2)

(640) A(5, 1) and B(0, −5)

ADVANCED ALGEBRA 1

Volume 1

Write the slope-intercept form of the equation of a straight line passing through the given point and parallel to the given straight line.

(641) $A(-2, 4)$,

parallel to the straight line $x = 0$

(642) $A(-3, -4)$,

parallel to the straight line $y = -3$

(643) $A(-2, -3)$,

parallel to the straight line $y = x - 4$

(644) $A(0, -5)$,

parallel to the straight line $y = -\dfrac{1}{2}x - 4$

(645) $A(2, 2)$,

parallel to the straight line $y = 3x + 1$

(646) $A(2, -5)$,

parallel to the straight line $y = -\dfrac{4}{7}x - 1$

(647) $A(-3, 0)$,

parallel to the straight line $y = -\dfrac{4}{3}x - 5$

(648) $A(1, -2)$,

parallel to the straight line $y = 2x + 2$

(649) $A(-5, 1)$,

parallel to the straight line $y = -\dfrac{6}{5}x + 1$

(650) $A(4, 4)$,

parallel to the straight line $y = \dfrac{5}{4}x + 5$

ADVANCED ALGEBRA 1

Volume 1

Write the slope-intercept form of the equation of a straight line passing through the given point and parallel to the given straight line.

(651) $A(-2, 3)$,
parallel to the straight line $y = -\dfrac{3}{2}x + 3$

(652) $A(2, 5)$,
parallel to the straight line $x = 0$

(653) $A(4, 1)$,
parallel to the straight line $y = -\dfrac{1}{4}x + 1$

(654) $A(2, -2)$,
parallel to the straight line $y = -\dfrac{3}{2}x - 2$

(655) $A(4, -3)$,
parallel to the straight line $y = -2x - 1$

(656) $A(2, -1)$,
parallel to the straight line $y = 2x + 1$

(657) $A(5, -2)$,
parallel to the straight line $y = -\dfrac{7}{5}x$

(658) $A(-3, 0)$,
parallel to the straight line $y = \dfrac{1}{3}x + 4$

(659) $A(-2, 0)$,
parallel to the straight line $y = x - 3$

(660) $A(2, 1)$,
parallel to the straight line $y = -x + 4$

Write the slope-intercept form of the equation of a straight line passing through the given point and parallel to the given straight line.

(661) A(2, 5),
parallel to the straight line $y = 4x + 1$

(662) A(3, −4),
parallel to the straight line $y = \frac{1}{3}x + 4$

(663) A(−2, 5),
parallel to the straight line $y = -\frac{3}{2}x + 4$

(664) A(5, 2),
parallel to the straight line $y = -x + 2$

(665) A(4, −2),
parallel to the straight line $y = \frac{3}{4}x + 2$

ADVANCED ALGEBRA 1

Volume 1

Write the point-slope form of the equation of a straight line passing through the given point and parallel to the given straight line.

(666) A(3, 5),
parallel to the point $y = \dfrac{7}{3}x - 5$

(667) A(−3, 2),
parallel to the point $y = -\dfrac{1}{3}x + 5$

(668) A(−5, −4),
parallel to the point $y = \dfrac{9}{5}x$

(669) A(−3, −1),
parallel to the point $y = \dfrac{3}{4}x - 1$

(670) A(−2, 5),
parallel to the point $y = 4x$

(671) A(0, 5),
parallel to the point $y = -2x - 3$

(672) A(4, 1),
parallel to the point $y = -\dfrac{1}{4}x + 1$

(673) A(−5, −4),
parallel to the point $y = \dfrac{1}{5}x + 1$

(674) A(5, 1),
parallel to the point $y = -\dfrac{3}{5}x + 3$

(675) A(2, −3),
parallel to the point $y = -\dfrac{3}{4}x - 1$

ADVANCED ALGEBRA 1

Volume 1

Write the point-slope form of the equation of a straight line passing through the given point and parallel to the given straight line.

(676) $A(-5, 5)$,
parallel to the point $y = -\dfrac{9}{5}x + 1$

(677) $A(1, 0)$,
parallel to the point $y = -4x + 2$

(678) $A(1, 5)$,
parallel to the point $y = 10x - 2$

(679) $A(4, 3)$,
parallel to the point $y = \dfrac{1}{2}x + 2$

(680) $A(-3, -5)$,
parallel to the point $y = \dfrac{2}{3}x + 5$

(681) $A(1, -4)$,
parallel to the point $y = -3x + 1$

(682) $A(3, -1)$,
parallel to the point $y = \dfrac{1}{3}x + 2$

(683) $A(2, 0)$,
parallel to the point $y = \dfrac{2}{3}x + 5$

(684) $A(4, -2)$,
parallel to the point $y = -\dfrac{5}{6}x$

(685) $A(-1, -5)$,
parallel to the point $y = 4x + 1$

ADVANCED ALGEBRA 1

Write the point-slope form of the equation of a straight line passing through the given point and parallel to the given straight line.

(686) A(−3, 5),
parallel to the point $y = -\frac{1}{3}x + 2$

(687) A(−4, 0),
parallel to the point $y = -\frac{3}{4}x - 2$

(688) A(1, 4),
parallel to the point $y = 6x - 3$

(689) A(5, 4),
parallel to the point $y = \frac{1}{5}x - 1$

(690) A(−4, −4),
parallel to the point $y = \frac{7}{4}x - 5$

ADVANCED ALGEBRA 1

Volume 1

Write the standard form of the equation of a straight line passing through the given point and parallel to the given straight line.

(691) A(3, −5),
parallel to the point $y = -x + 1$

(692) A(4, 3),
parallel to the point $y = \frac{1}{2}x + 5$

(693) A(5, 0),
parallel to the point $y = \frac{1}{3}x + 1$

(694) A(−4, −1),
parallel to the point $y = \frac{3}{2}x - 2$

(695) A(5, −1),
parallel to the point $y = \frac{1}{5}x + 3$

(696) A(−5, 1),
parallel to the point $y = -\frac{1}{10}x - 1$

(697) A(5, 3),
parallel to the point $y = \frac{6}{5}x + 5$

(698) A(−5, 2),
parallel to the point $y = -3$

(699) A(3, −2),
parallel to the point $y = -\frac{5}{3}x - 3$

(700) A(−2, 5),
parallel to the point $y = -1$

ADVANCED ALGEBRA 1

Volume 1

Write the standard form of the equation of a straight line passing through the given point and parallel to the given straight line.

(701) A(5, −2),
   parallel to the point $y = -\dfrac{2}{5}x - 5$

(702) A(4, −3),
   parallel to the point $y = -\dfrac{3}{2}x + 4$

(703) A(−1, 4),
   parallel to the point $y = -2x - 4$

(704) A(4, −1),
   parallel to the point $y = -4x + 4$

(705) A(3, 0),
   parallel to the point $x = 0$

(706) A(2, 2),
   parallel to the point $y = -x - 3$

(707) A(−5, −4),
   parallel to the point $y = \dfrac{8}{7}x + 3$

(708) A(−5, −2),
   parallel to the point $y = x - 2$

(709) A(−3, −3),
   parallel to the point $y = \dfrac{8}{3}x + 2$

(710) A(−1, 0),
   parallel to the point $y = x - 5$

ADVANCED ALGEBRA 1

Volume 1

Write the standard form of the equation of a straight line passing through the given point and parallel to the given straight line.

(711) A(3, 3),
parallel to the point $y = \dfrac{8}{3}x + 4$

(712) A(5, −1),
parallel to the point $y = -\dfrac{3}{5}x$

(713) A(2, −1),
parallel to the point $y = -3x - 5$

(714) A(3, 1),
parallel to the point $y = x + 5$

(715) A(−5, 3),
parallel to the point $y = -\dfrac{6}{5}x - 1$

(716) A(−4, −4),
perpendicular to the point $y = -x + 3$

(717) A(3, 5),
perpendicular to the point $y = -\dfrac{1}{3}x + 1$

(718) A(3, −3),
perpendicular to the point $y = x - 3$

(719) A(−3, 2),
perpendicular to the point $y = \dfrac{3}{7}x + 2$

(720) A(4, −5),
perpendicular to the point $y = \dfrac{1}{2}x + 4$

ADVANCED ALGEBRA 1

Volume 1

Write the standard form of the equation of a straight line passing through the given point and parallel to the given straight line.

(721) A(−1, −3),
perpendicular to the point $y = -\frac{1}{2}x + 2$

(722) A(4, 1),
perpendicular to the point $y = -\frac{4}{5}x - 5$

(723) A(5, 3),
perpendicular to the point $y = \frac{5}{2}x + 1$

(724) A(3, 3),
perpendicular to the point $y = -3x + 2$

(725) A(2, 4),
perpendicular to the point $y = -\frac{2}{7}x + 3$

(726) A(4, 4),
perpendicular to the point $y = -2x - 2$

(727) A(−1, 2),
perpendicular to the point $y = -\frac{1}{5}x - 1$

(728) A(4, −4),
perpendicular to the point $y = -\frac{1}{6}x - 2$

(729) A(5, 0),
perpendicular to the point $y = \frac{2}{3}x - 3$

(730) A(−1, 2),
perpendicular to the point $y = \frac{1}{6}x - 5$

Write the standard form of the equation of a straight line passing through the given point and parallel to the given straight line.

(731) A(−1, −1),
perpendicular to the point $y = x − 2$

(732) A(−4, 4),
perpendicular to the point $y = \dfrac{2}{3}x − 5$

(733) A(−4, −3),
perpendicular to the point $x = 0$

(734) A(−5, −1),
perpendicular to the point $y = \dfrac{5}{3}x − 3$

(735) A(−4, 0),
perpendicular to the point $y = −\dfrac{4}{3}x + 2$

(736) A(−1, 0),
perpendicular to the point $y = \dfrac{2}{3}x + 4$

(737) A(1, 4),
perpendicular to the point $y = x + 3$

(738) A(−1, −2),
perpendicular to the point $y = −\dfrac{1}{7}x − 4$

(739) A(0, 0),
perpendicular to the point $y = −4$

(740) A(−2, −5),
perpendicular to the point $y = −1$

ADVANCED ALGEBRA 1

Write the slope-intercept form of the equation of a straight line passing through the given point and perpendicular to the given straight line.

(741) A(4, −3),
  perpendicular to the point $y = -4x + 1$

(742) A(2, 4),
  perpendicular to the point $y = \frac{1}{2}x - 4$

(743) A(1, −1),
  perpendicular to the point $y = -\frac{1}{4}x + 1$

(744) A(1, 5),
  perpendicular to the point $y = -\frac{1}{9}x + 5$

(745) A(3, −3),
  perpendicular to the point $y = -3x - 3$

(746) A(5, −3),
  perpendicular to the point $y = \frac{5}{3}x + 4$

(747) A(3, −1),
  perpendicular to the point $y = -\frac{3}{4}x - 2$

(748) A(3, −3),
  perpendicular to the point $y = x$

(749) A(−5, 4),
  perpendicular to the point $y = \frac{5}{8}x - 3$

(750) A(−1, 5),
  perpendicular to the point $y = \frac{1}{2}x - 5$

ADVANCED ALGEBRA 1

Volume 1

Write the slope-intercept form of the equation of a straight line passing through the given point and perpendicular to the given straight line.

(751) A(5, 4),
perpendicular to the point $y = -5x + 4$

(752) A(1, 3),
perpendicular to the point $y = -\dfrac{1}{4}x - 5$

(753) A(1, 0),
perpendicular to the point $y = -\dfrac{1}{4}x - 4$

(754) A(−3, 4),
perpendicular to the point $y = \dfrac{3}{5}x + 3$

(755) A(3, 0),
perpendicular to the point $y = \dfrac{2}{5}x$

(756) A(5, −1),
perpendicular to the point $y = 5x + 1$

(757) A(−4, −5),
perpendicular to the point $y = -\dfrac{4}{9}x + 4$

(758) A(2, 5),
perpendicular to the point $y = -\dfrac{2}{9}x - 1$

(759) A(−4, −5),
perpendicular to the point $y = -\dfrac{5}{9}x + 4$

(760) A(5, 0),
perpendicular to the point $y = -\dfrac{5}{3}x + 2$

Write the slope-intercept form of the equation of a straight line passing through the given point and perpendicular to the given straight line.

(761) A(−5, 4),

perpendicular to the point $y = -5x - 3$

(762) A(−2, 0),

perpendicular to the point $y = \frac{3}{5}x - 3$

(763) A(4, 1),

perpendicular to the point $y = 2x - 2$

(764) A(−1, −2),

perpendicular to the point $y = -2x - 3$

(765) A(−4, 1),

perpendicular to the point $y = -4x - 5$

ADVANCED ALGEBRA 1

Volume 1

Write the point-slope form of the equation of a straight line passing through the given point and perpendicular to the given straight line.

(766) A(1, 4),
perpendicular to the point $y = -\dfrac{1}{5}x + 4$

(767) A(−3, −4),
perpendicular to the point $y = x - 1$

(768) A(−3, −3),
perpendicular to the point $y = -\dfrac{1}{2}x + 3$

(769) A(2, 3),
perpendicular to the point $y = 2x - 2$

(770) A(−2, 5),
perpendicular to the point $y = \dfrac{1}{3}x - 2$

(771) A(−2, −2),
perpendicular to the point $x = 0$

(772) A(−4, −3),
perpendicular to the point $y = \dfrac{1}{7}x$

(773) A(−2, −1),
perpendicular to the point $y = -2x$

(774) A(−2, −2),
perpendicular to the point $y = \dfrac{1}{3}x - 2$

(775) A(4, 2),
perpendicular to the point $y = \dfrac{3}{2}x - 3$

ADVANCED ALGEBRA 1

Volume 1

Write the point-slope form of the equation of a straight line passing through the given point and perpendicular to the given straight line.

(776) $A(-5, 3)$,

perpendicular to the point $y = \dfrac{2}{5}x + 4$

(777) $A(1, 1)$,

perpendicular to the point $y = 3x - 2$

(778) $A(0, 0)$,

perpendicular to the point $y = x$

(779) $A(-5, 1)$,

perpendicular to the point $y = -\dfrac{5}{4}x + 3$

(780) $A(-1, -5)$,

perpendicular to the point $y = -\dfrac{1}{8}x - 5$

(781) $A(-5, 3)$,

perpendicular to the point $y = x + 3$

(782) $A(-4, 1)$,

perpendicular to the point $y = 2x$

(783) $A(-5, 3)$,

perpendicular to the point $y = 2x - 4$

(784) $A(-5, -1)$,

perpendicular to the point $y = -\dfrac{1}{3}x - 1$

(785) $A(3, -4)$,

perpendicular to the point $y = 4$

Write the point-slope form of the equation of a straight line passing through the given point and perpendicular to the given straight line.

(786) A(−1, 3),
perpendicular to the point $y = -\frac{1}{2}x - 2$

(787) A(2, 5),
perpendicular to the point $y = -\frac{1}{2}x - 4$

(788) A(4, 5),
perpendicular to the point $y = -x - 4$

(789) A(−2, −5),
perpendicular to the point $y = -\frac{1}{2}x - 5$

(790) A(−2, 0),
perpendicular to the point $x = 0$

Plot the graph for each linear inequality as given below.

(791) $x - y < 0$

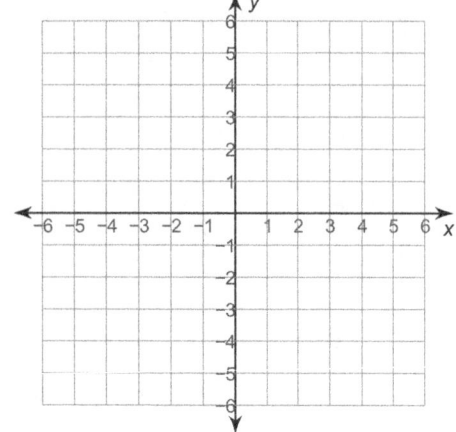

(792) $7x + y \geq -3$

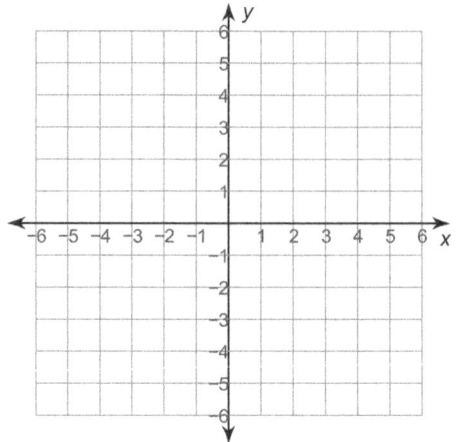

(793) $x + 4y > 0$

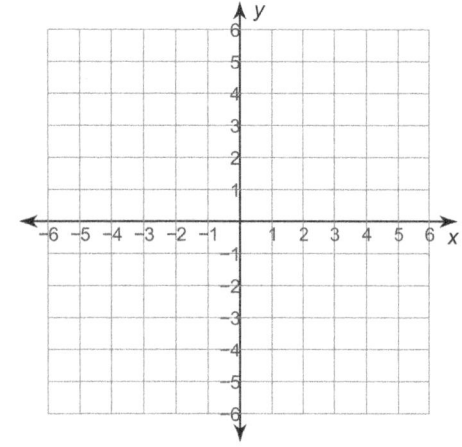

(794) $2x - 3y \geq -6$

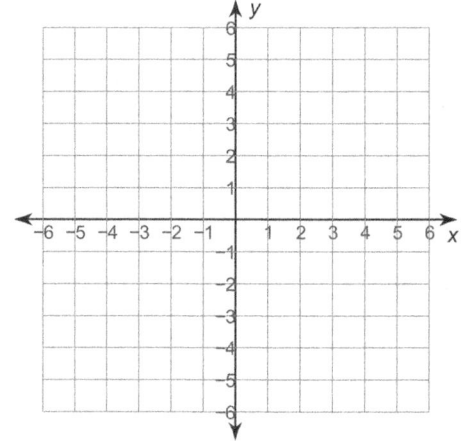

Plot the graph for each linear inequality as given below.

(795) $x + 4y \geq -20$

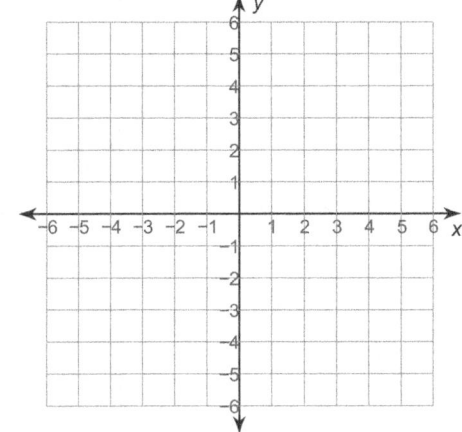

(796) $x - 5y \geq 15$

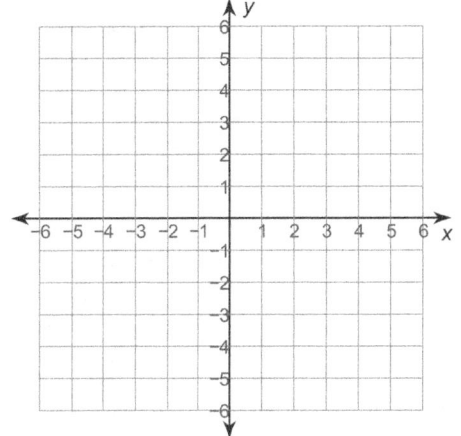

(797) $5x + 2y \leq 0$

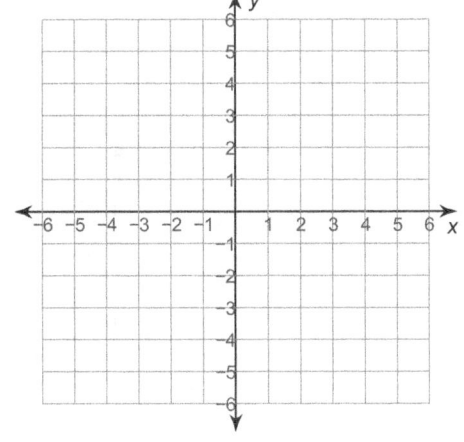

(798) $10x + y \leq 5$

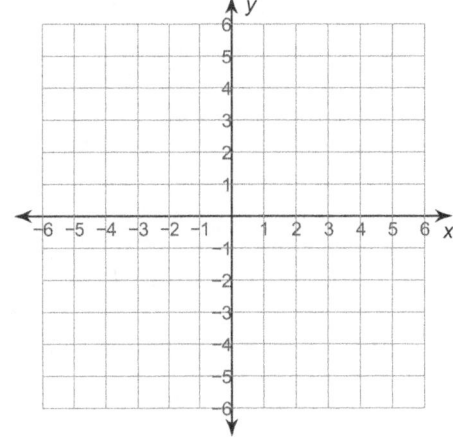

Plot the graph for each linear inequality as given below.

(799) $9x + 4y \geq 16$

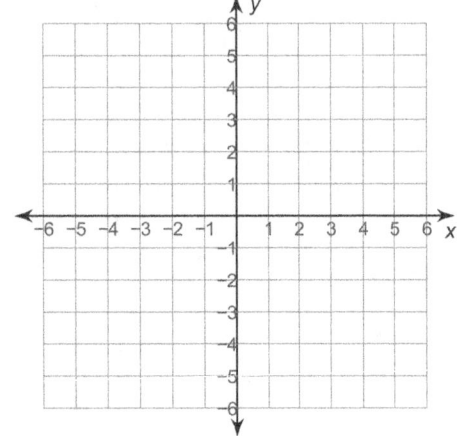

(800) $x + y \geq -1$

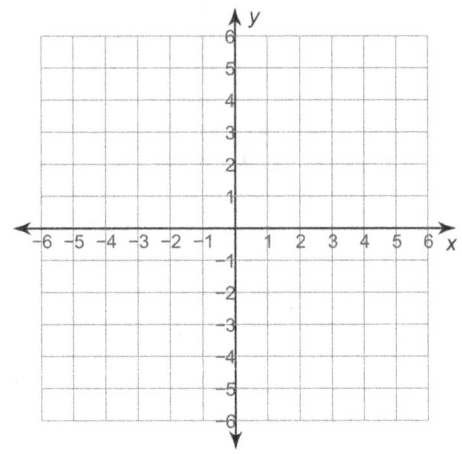

(801) $x > -2$

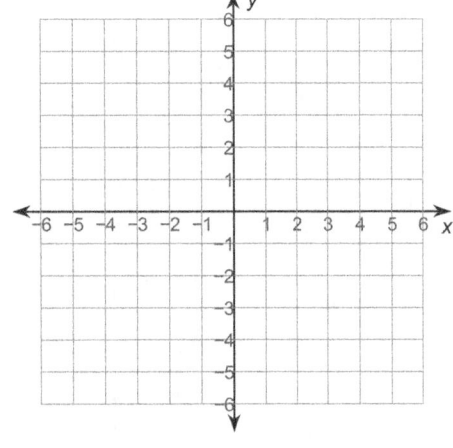

(802) $5x + y \leq 2$

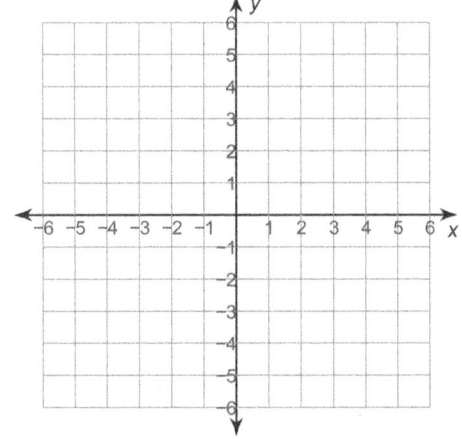

Plot the graph for each linear inequality as given below.

(803) $3x + y \leq -2$

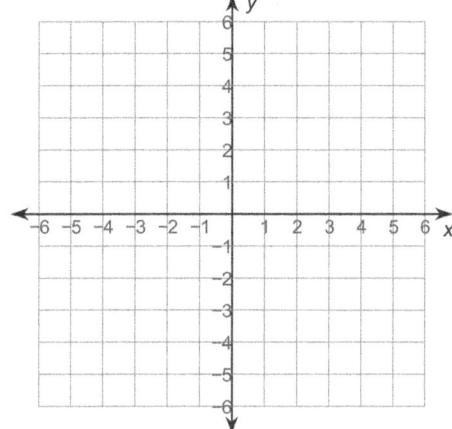

(804) $2x + y < -5$

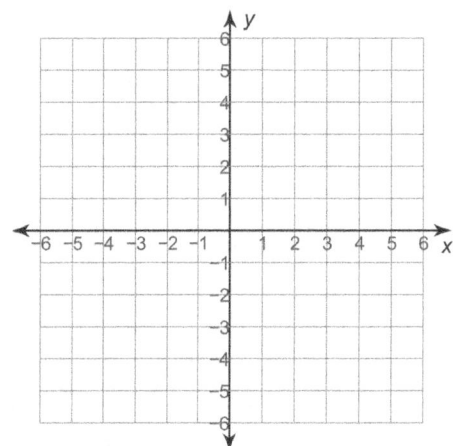

(805) $x + y < -4$

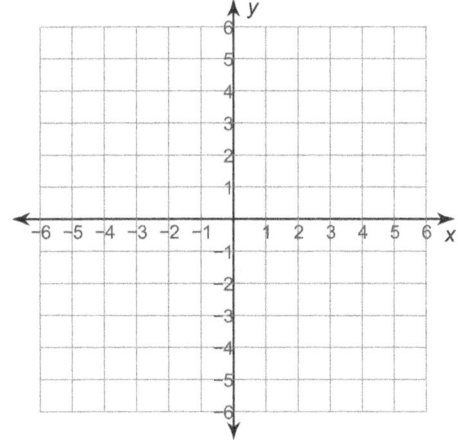

(806) $5x + 2y \leq 2$

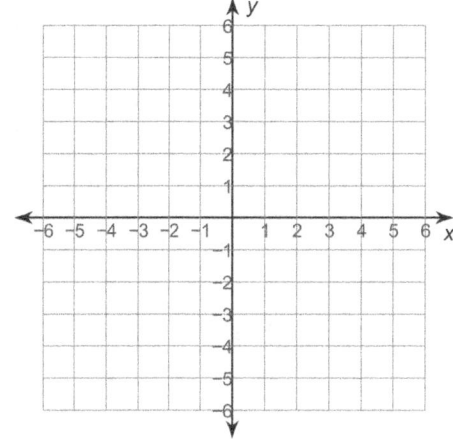

Plot the graph for each linear inequality as given below.

(807) $x - 4y < 4$

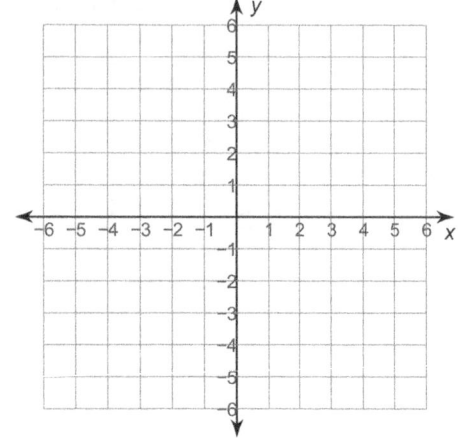

(808) $6x - y \geq 3$

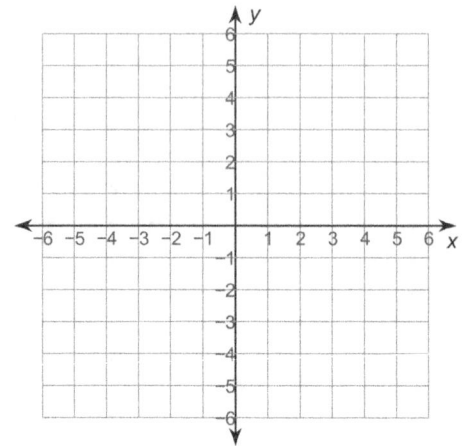

(809) $4x - y < 0$

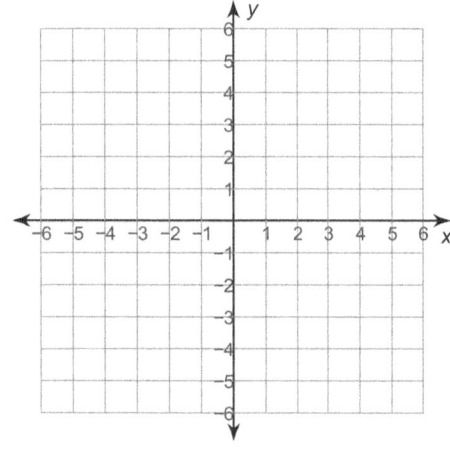

(810) $x + y \geq 2$

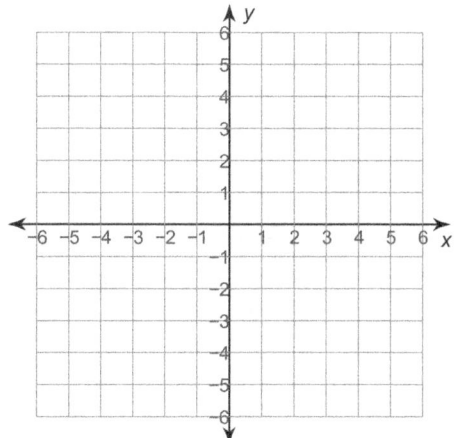

Plot the graph for each linear inequality as given below.

(811) $2x - 3y > -6$

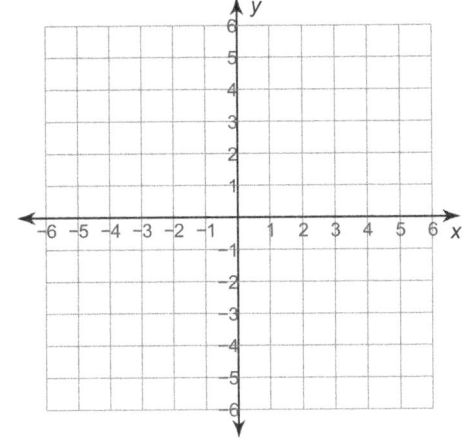

(812) $y \leq -5$

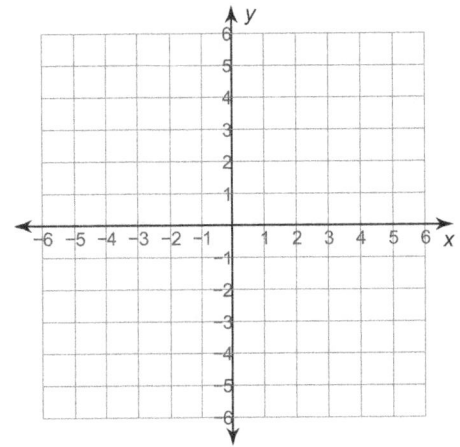

(813) $9x - 4y > 20$

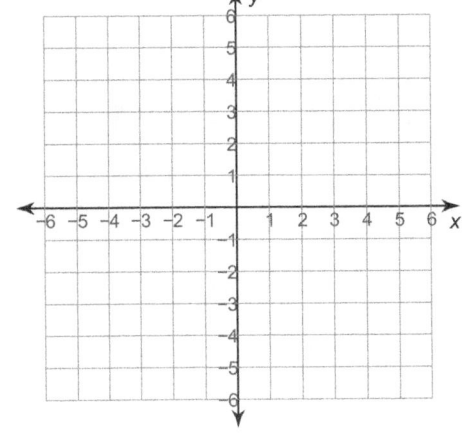

(814) $x - y < 1$

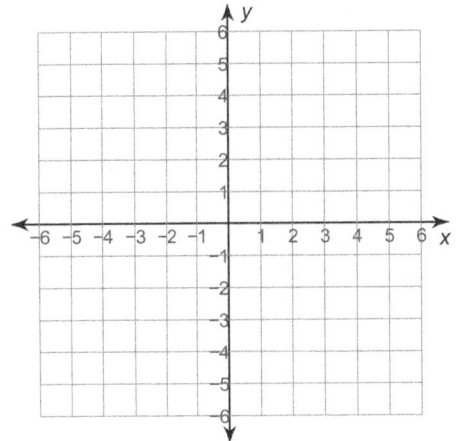

Plot the graph for each linear inequality as given below.

(815) $x - y \le -3$

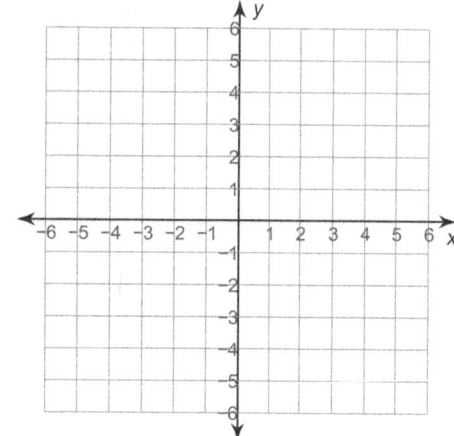

(816) $5x + y < 5$

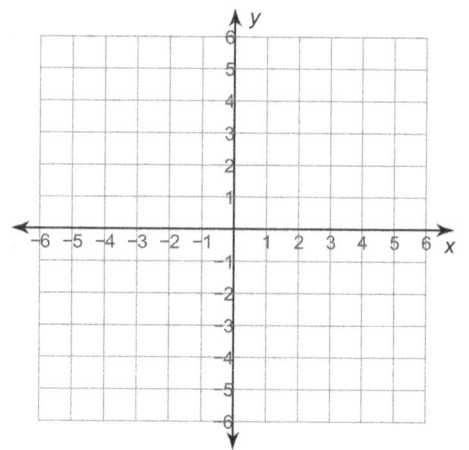

(817) $3x - 2y < -6$

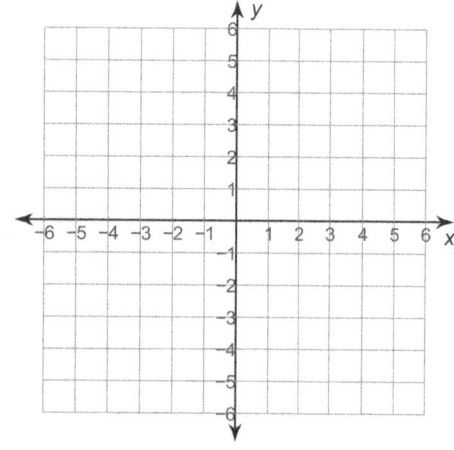

(818) $5x - 4y \ge 0$

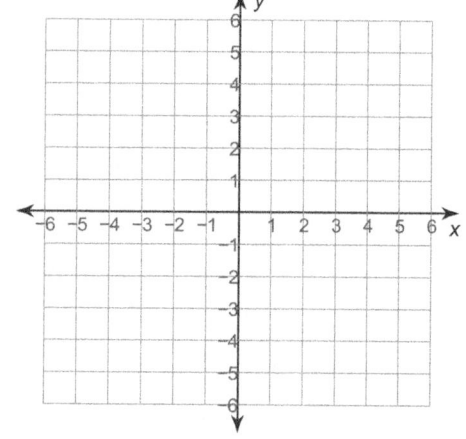

ADVANCED ALGEBRA 1

Volume 1

Plot the graph for each linear inequality as given below.

(819) $4x + y \geq 4$

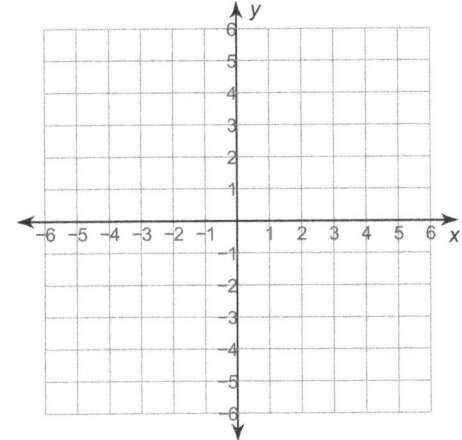

(820) $y < -4$

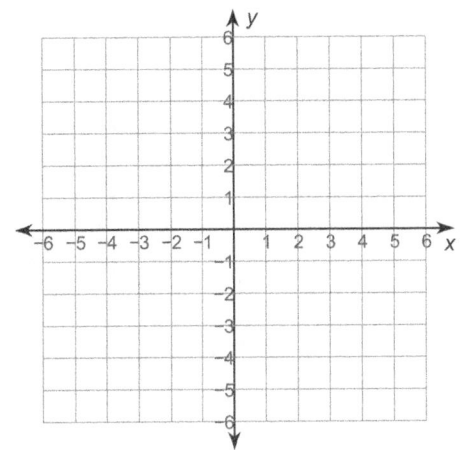

(821) $3x + y \geq -2$

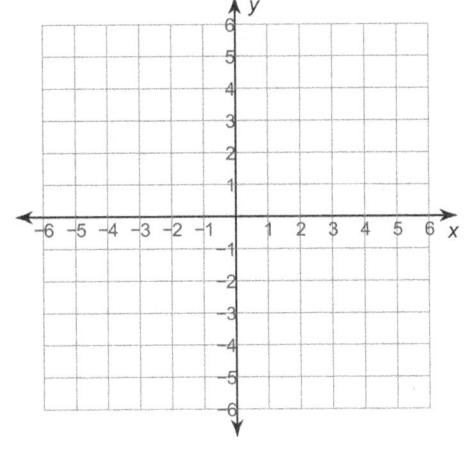

(822) $2x + y > 0$

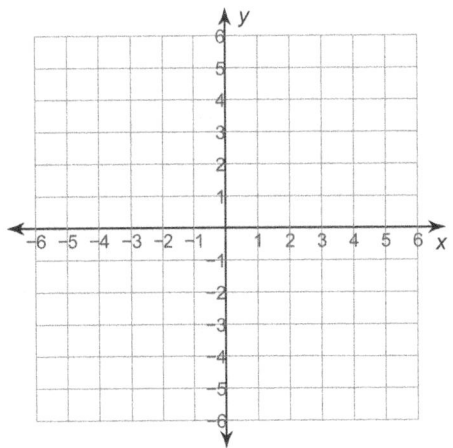

Plot the graph for each linear inequality as given below.

(823) $4x + 3y > 6$

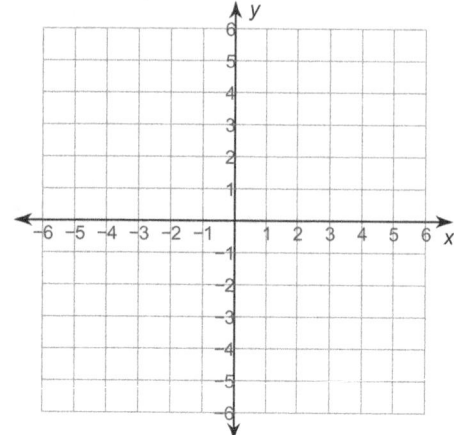

(824) $x + 4y \le -8$

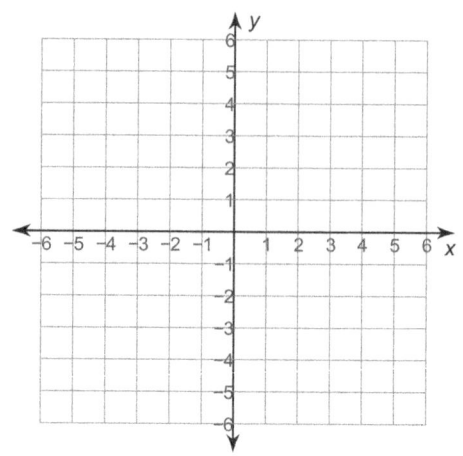

(825) $6x - 5y > 10$

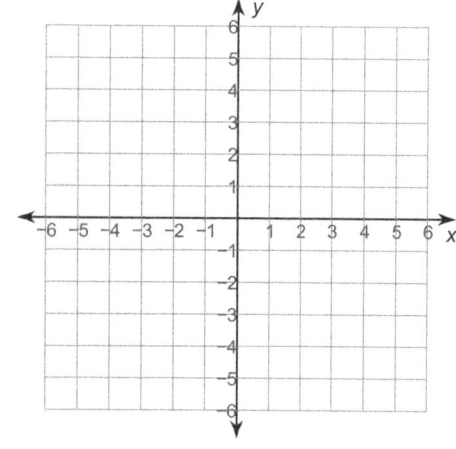

(826) $7x + 3y \le 12$

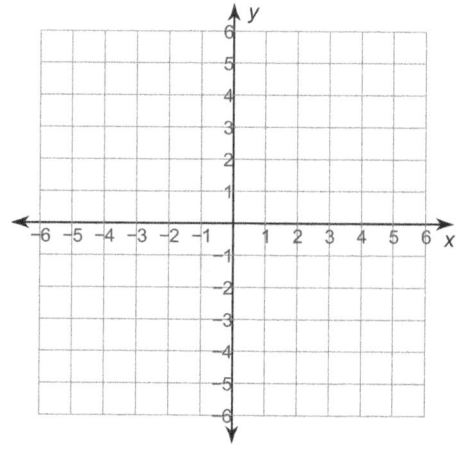

Plot the graph for each linear inequality as given below.

(827) $2x - 3y \geq 6$

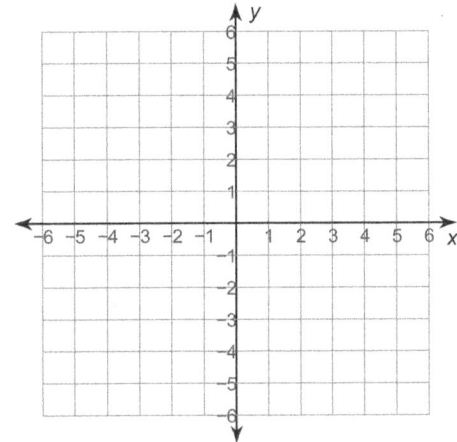

(828) $6x - 5y < -10$

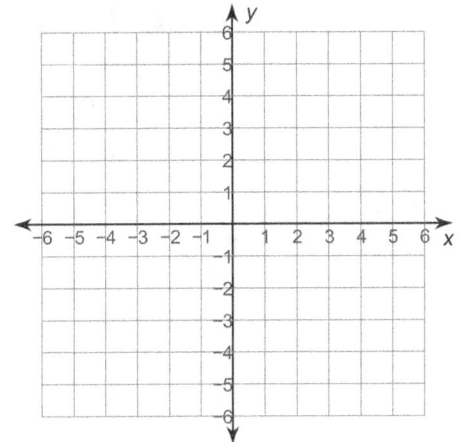

(829) $2x - 3y > -3$

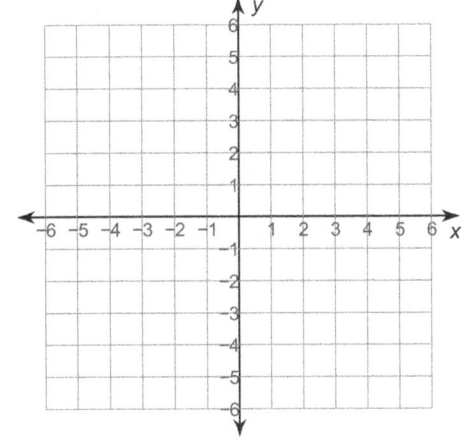

(830) $5x - 4y \leq -20$

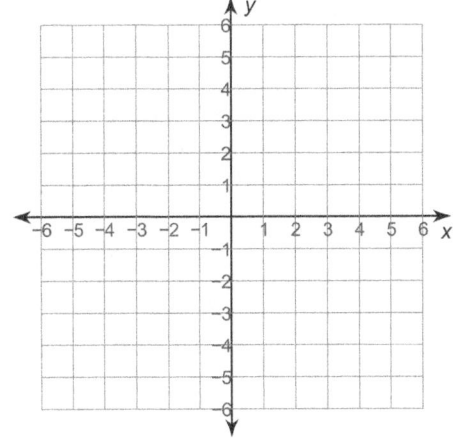

Plot the graph for each linear inequality as given below.

(831) $2x + 3y < -12$

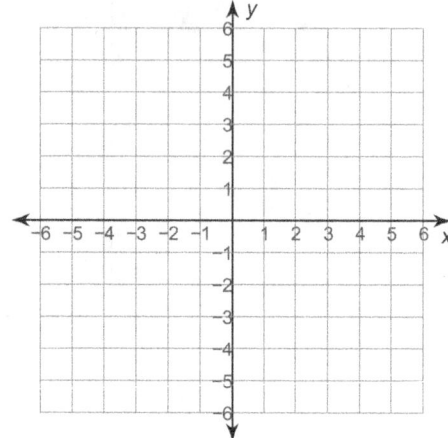

(832) $3x + 5y \leq 0$

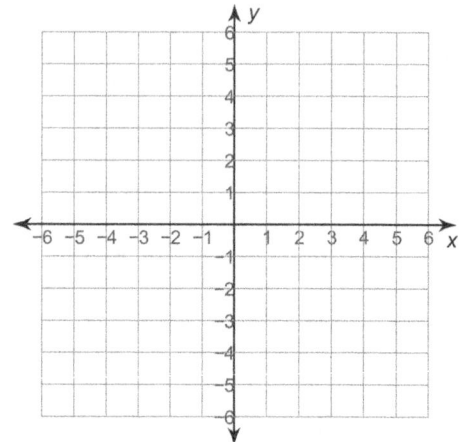

(833) $4x - 3y > 3$

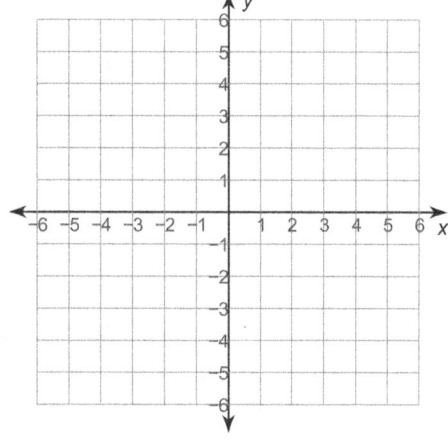

(834) $x > 3$

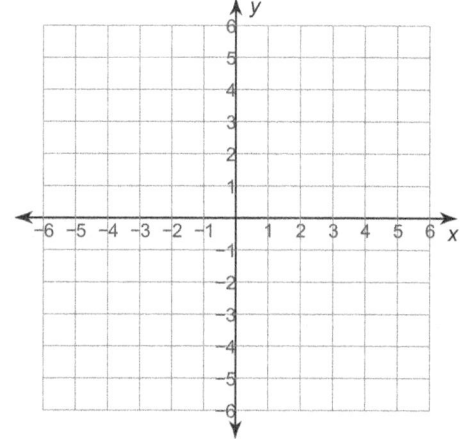

Plot the graph for each linear inequality as given below.

(835) $2x + y \leq -4$

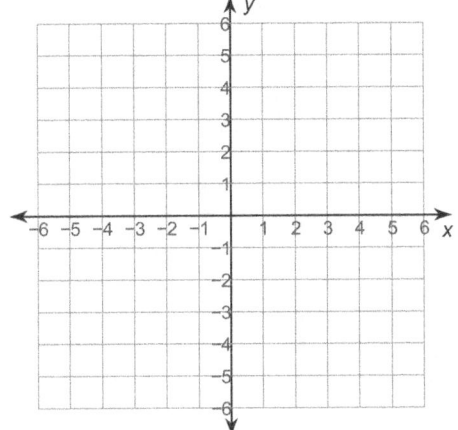

(836) $2x + y < 4$

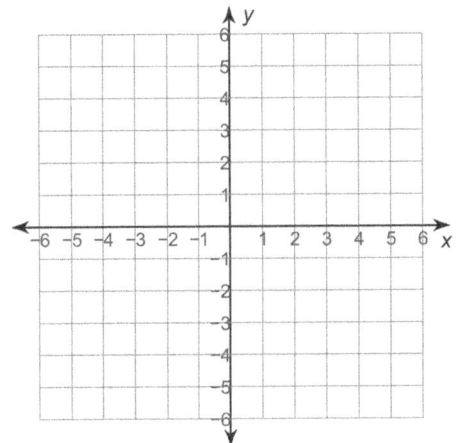

(837) $3x + 4y \leq -8$

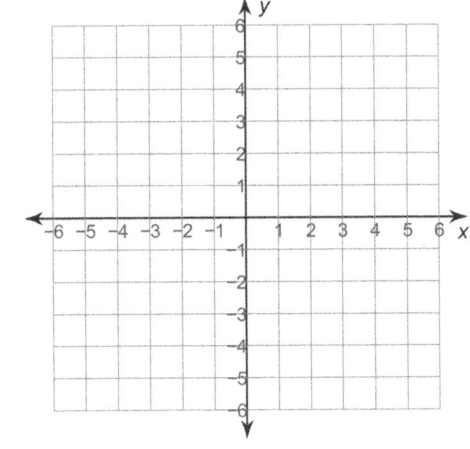

(838) $x + y \leq 5$

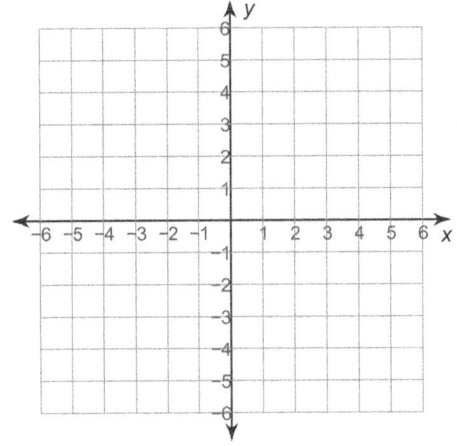

Plot the graph for each linear inequality as given below.

(839) $x - y \leq 3$

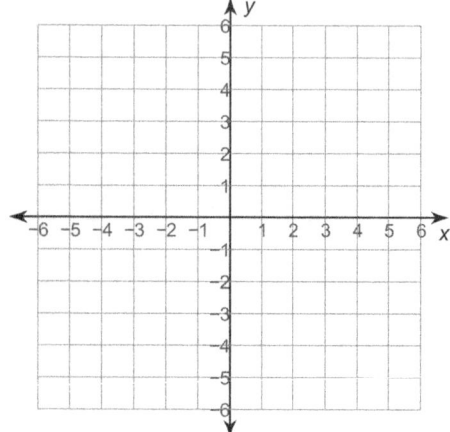

(840) $x \geq 1$

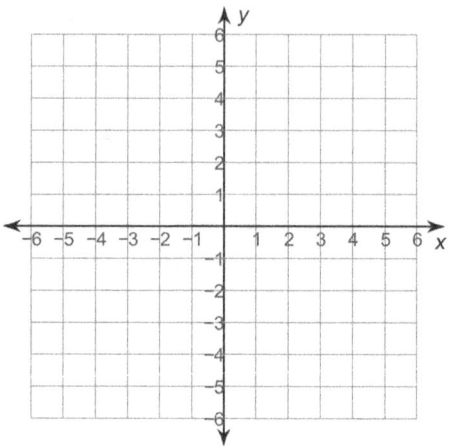

(841) $y > \dfrac{3}{2}x - 3$

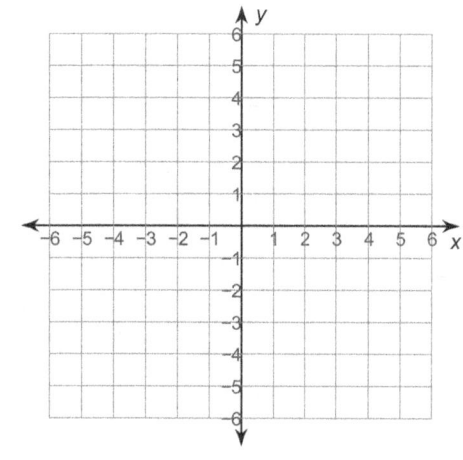

(842) $y > -3x + 2$

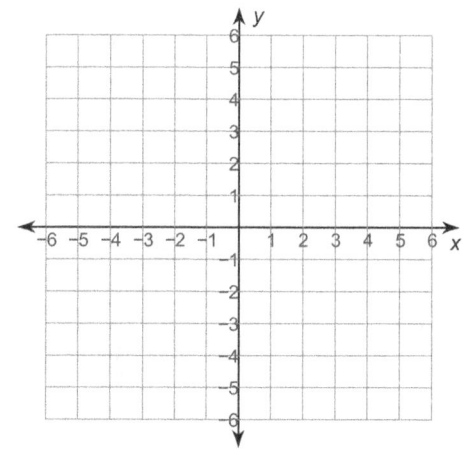

ADVANCED ALGEBRA 1

Volume 1

Plot the graph for each linear inequality as given below.

(843) $y < -5x$

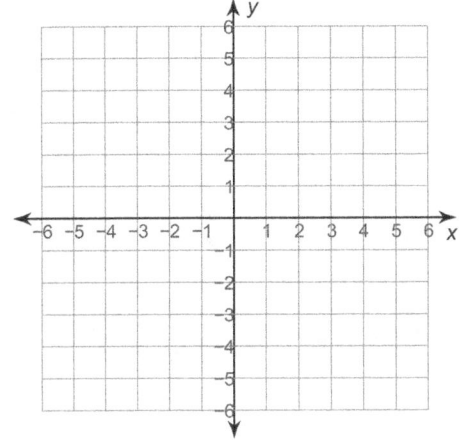

(844) $y < 2x - 1$

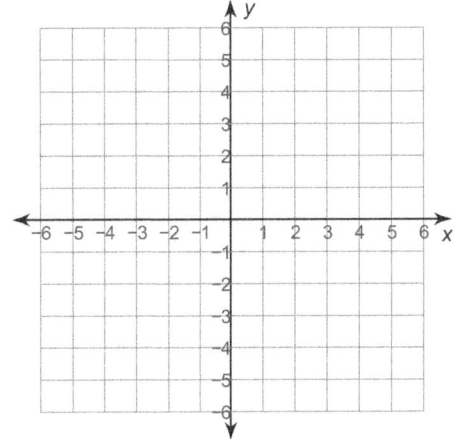

(845) $y < -x - 3$

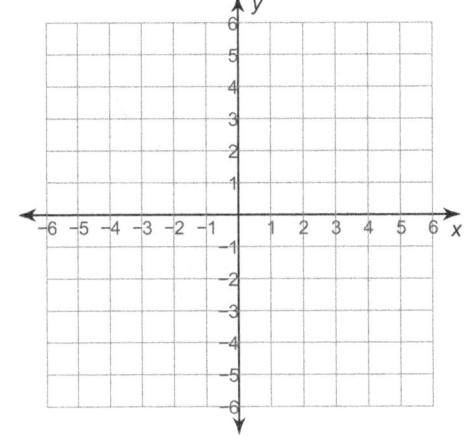

(846) $x > -2$

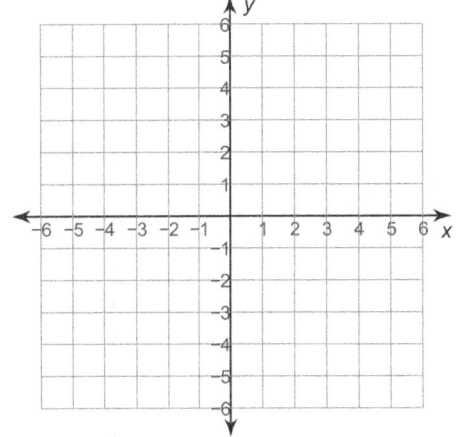

Plot the graph for each linear inequality as given below.

(847) $y < -2x + 1$

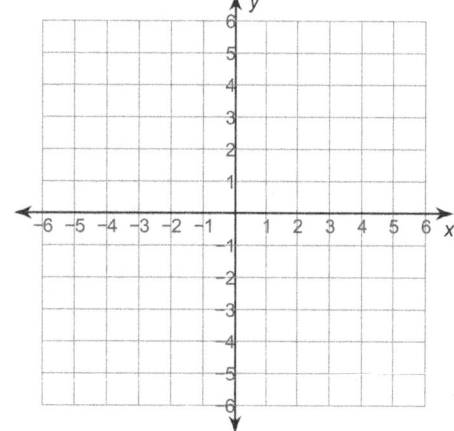

(848) $y > 2x - 5$

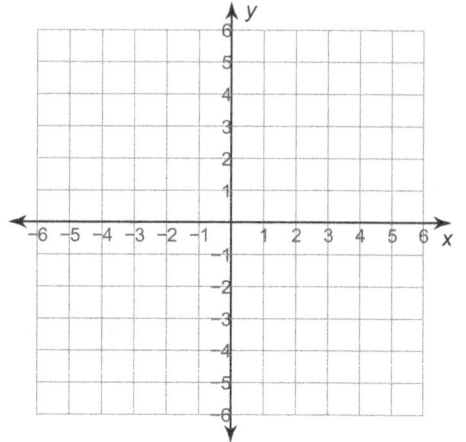

(849) $y > \dfrac{5}{2}x - 4$

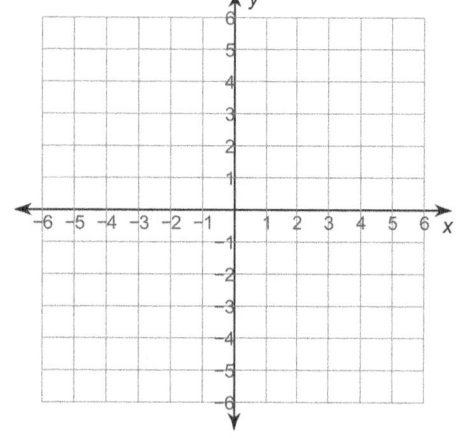

(850) $y < \dfrac{1}{2}x - 2$

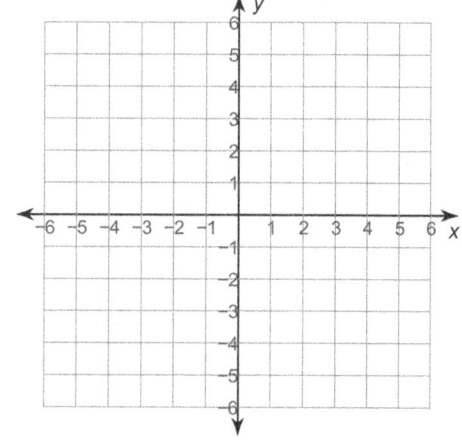

Plot the graph for each linear inequality as given below.

(851) $y > \dfrac{3}{2}x - 2$

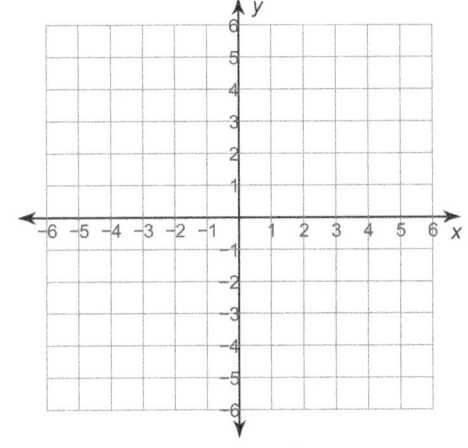

(852) $x > -1$

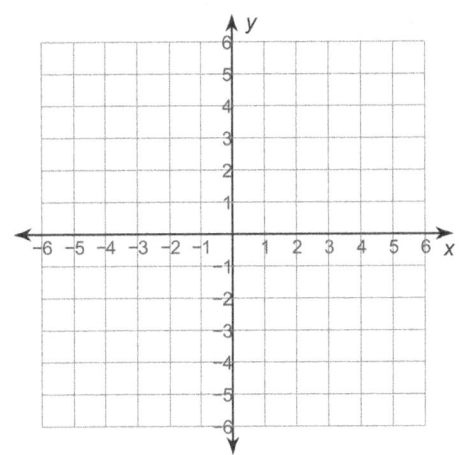

(853) $y < -2$

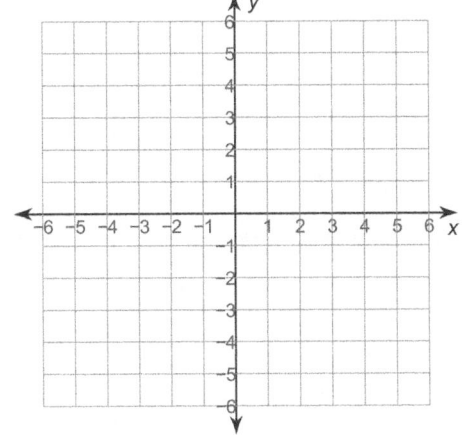

(854) $y < 4x - 4$

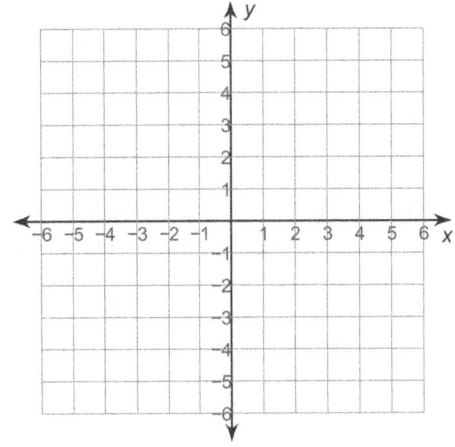

Plot the graph for each linear inequality as given below.

(855) $y < \dfrac{9}{4}x - 4$

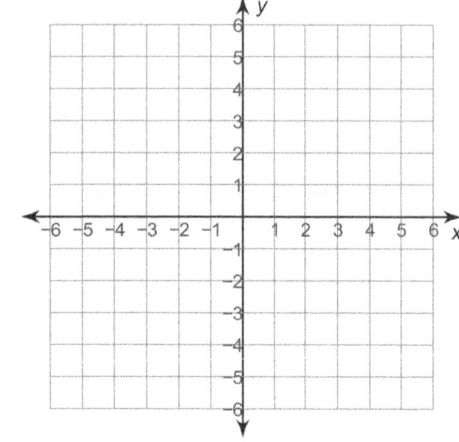

(856) $x \geq -3$

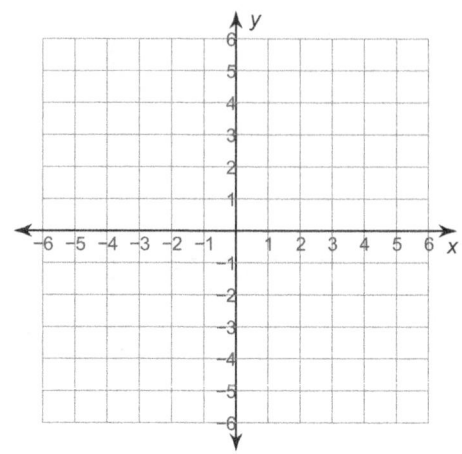

(857) $y \geq \dfrac{7}{5}x + 3$

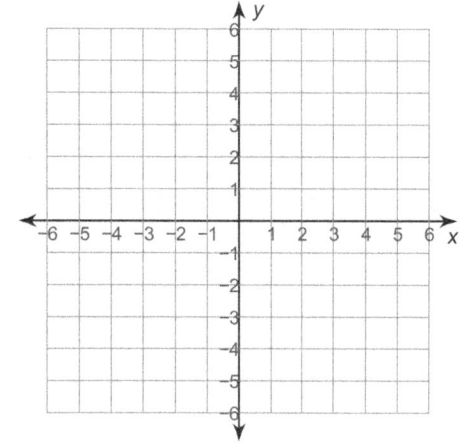

(858) $y \leq \dfrac{1}{2}x + 5$

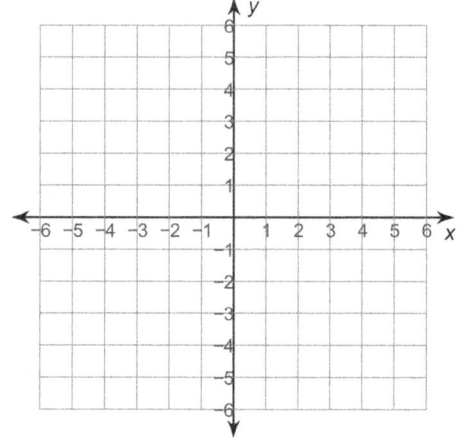

Plot the graph for each linear inequality as given below.

(859) $y \geq -2x + 4$

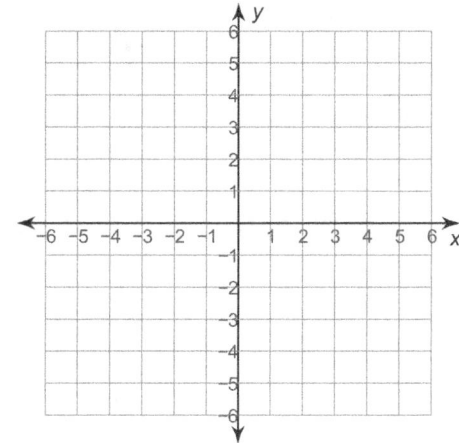

(860) $y > \dfrac{1}{3}x + 5$

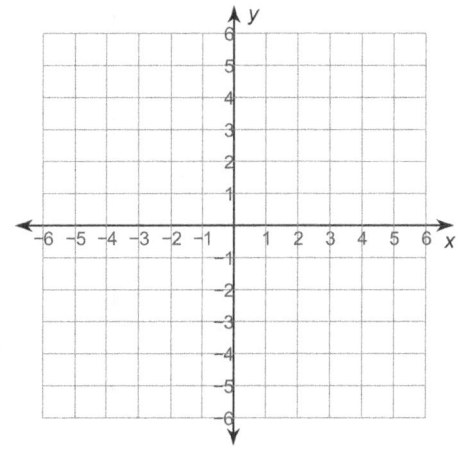

(861) $y < 3x - 4$

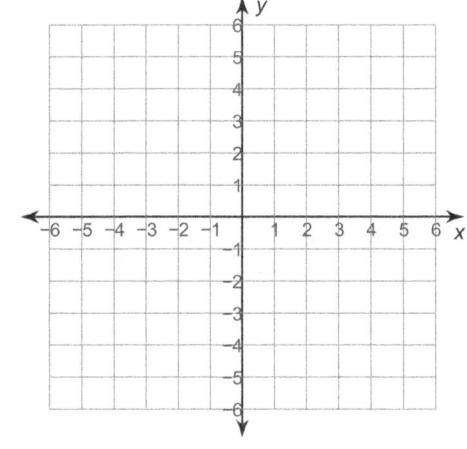

(862) $y < -\dfrac{2}{5}x + 4$

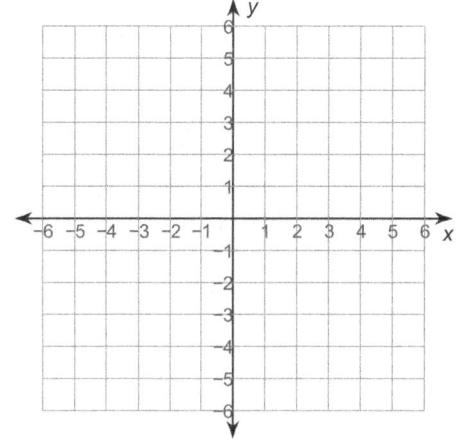

Plot the graph for each linear inequality as given below.

(863) $y < 1$

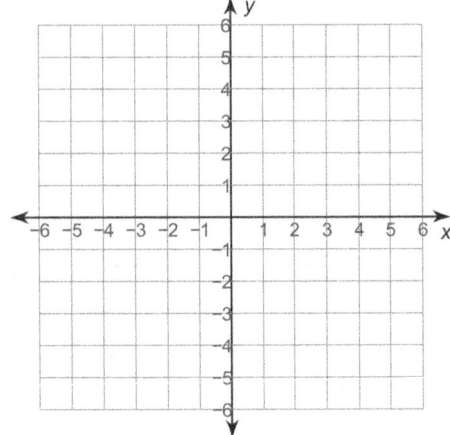

(864) $y < x + 2$

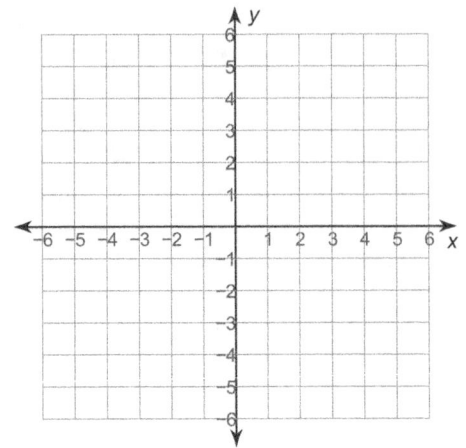

(865) $y < \dfrac{3}{2}x + 2$

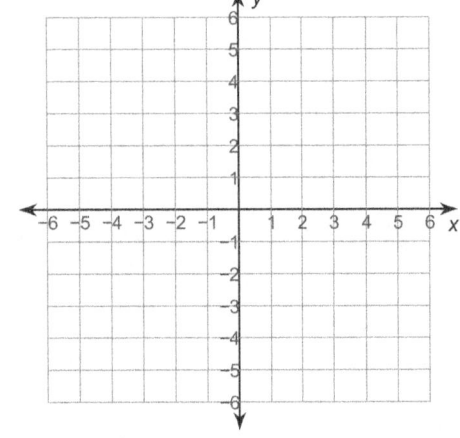

(866) $y < -\dfrac{7}{3}x - 5$

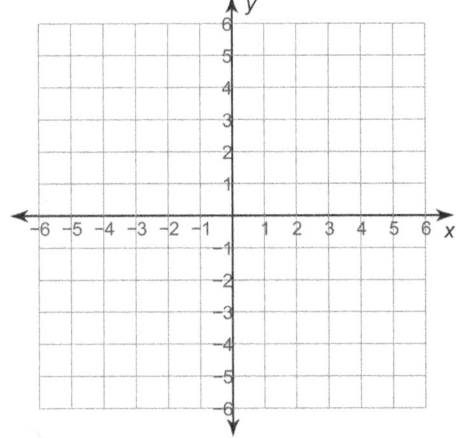

Plot the graph for each linear inequality as given below.

(867) $y < -\dfrac{1}{3}x - 1$

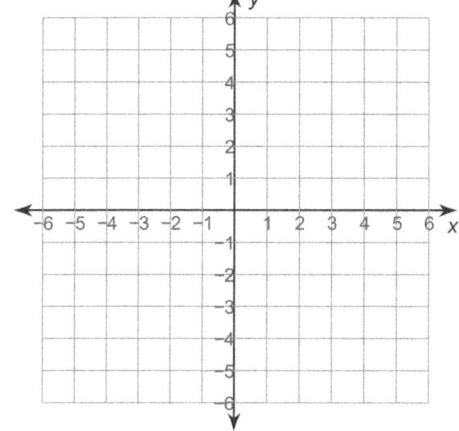

(868) $y < \dfrac{1}{3}x - 5$

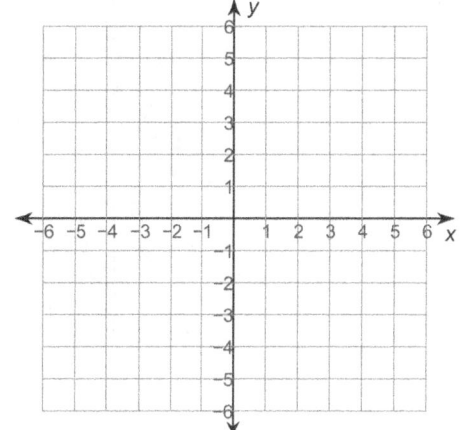

(869) $y \le \dfrac{1}{2}x - 2$

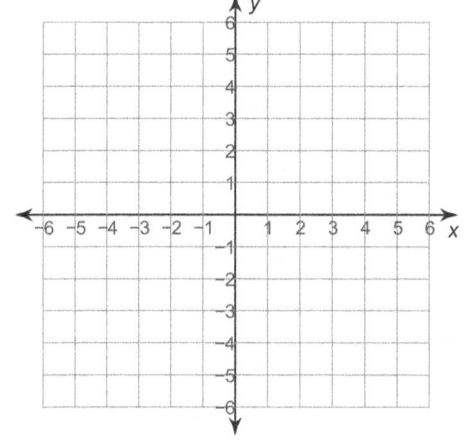

(870) $y < -\dfrac{3}{2}x - 3$

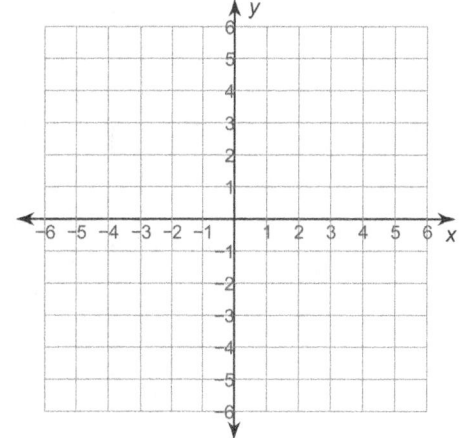

Plot the graph for each linear inequality as given below.

(871) $y \geq -\dfrac{3}{2}x + 1$

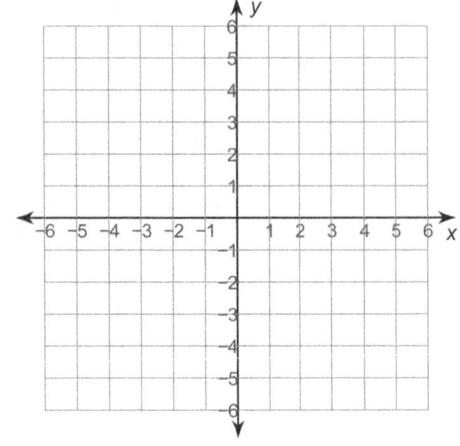

(872) $y \geq x - 1$

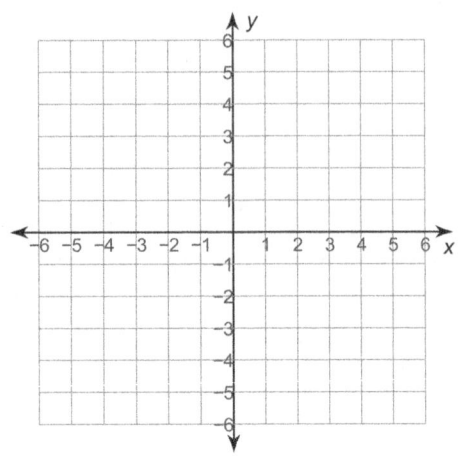

(873) $y \geq \dfrac{4}{3}x - 4$

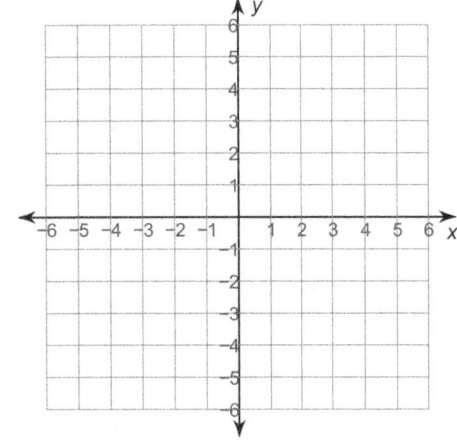

(874) $y \leq -\dfrac{7}{2}x + 2$

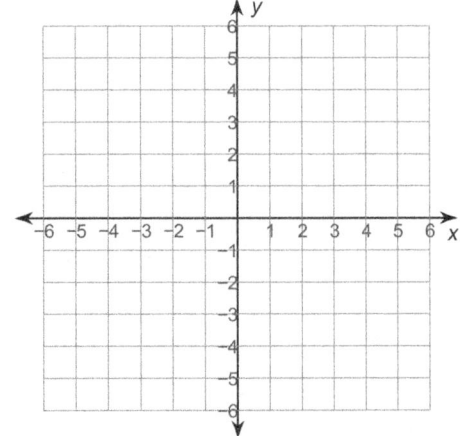

Plot the graph for each linear inequality as given below.

(875) $y < -\dfrac{9}{4}x - 5$

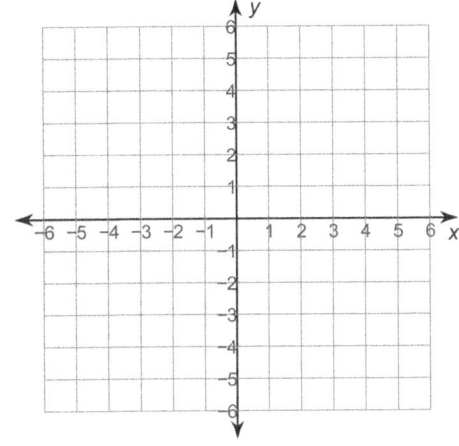

(876) $y \le 2x - 2$

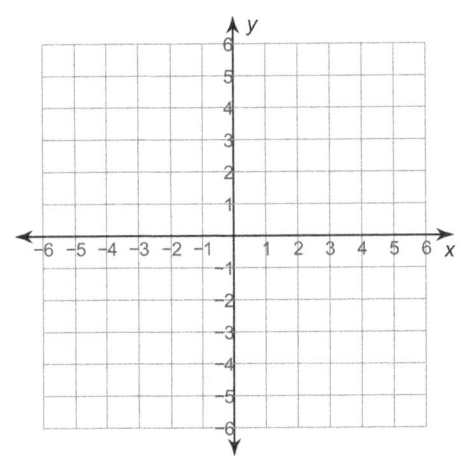

(877) $x < -3$

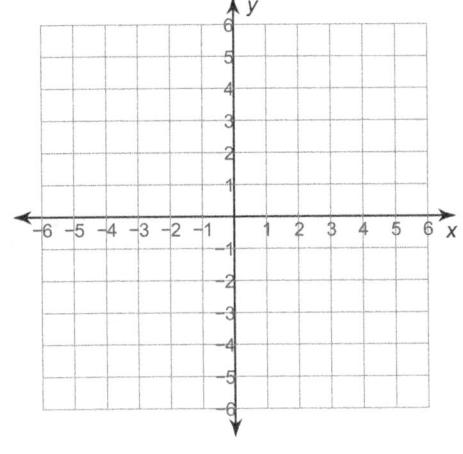

(878) $y < -x - 4$

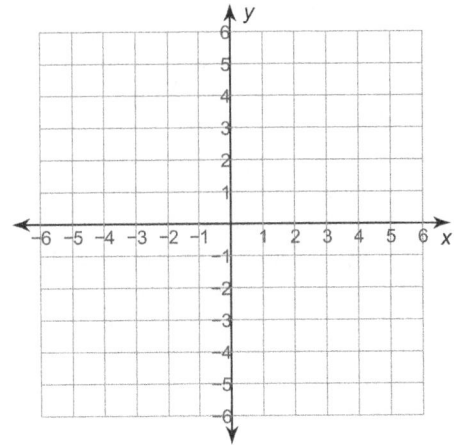

Plot the graph for each linear inequality as given below.

(879) $y > x$

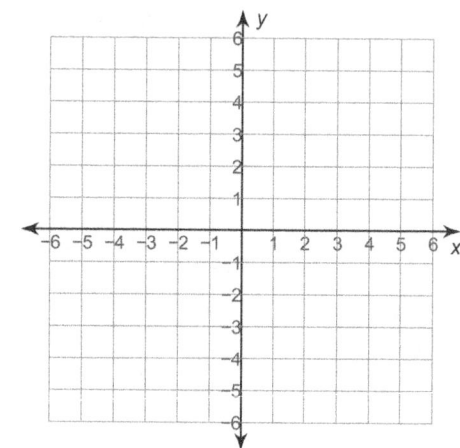

(880) $y > -\dfrac{1}{2}x + 5$

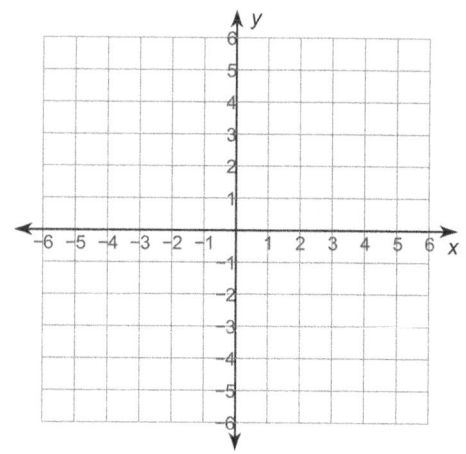

(881) $y > \dfrac{3}{4}x + 2$

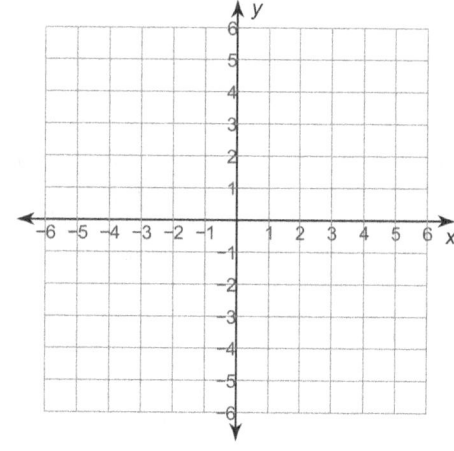

(882) $y < 3x$

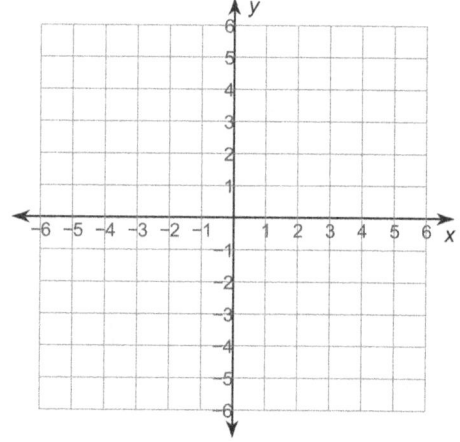

Plot the graph for each linear inequality as given below.

(883) $y > x - 5$

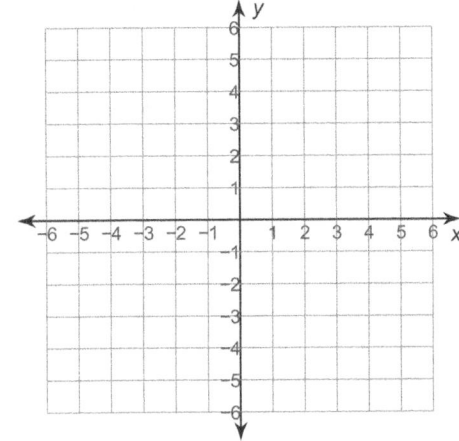

(884) $y > -\dfrac{5}{2}x - 1$

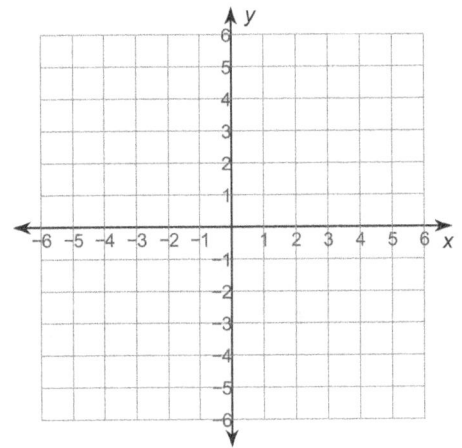

(885) $y \leq \dfrac{1}{4}x + 5$

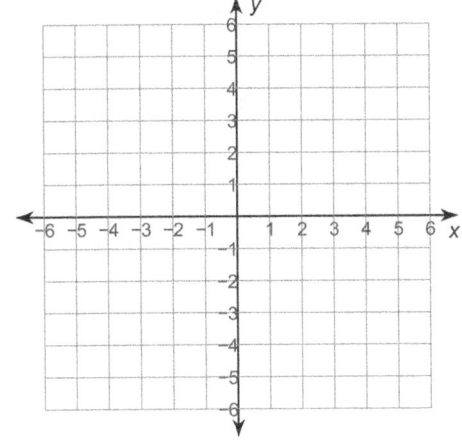

(886) $y > \dfrac{2}{5}x + 4$

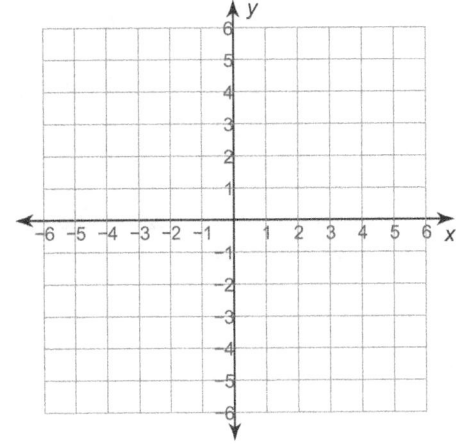

Plot the graph for each linear inequality as given below.

(887) $y > -x - 2$

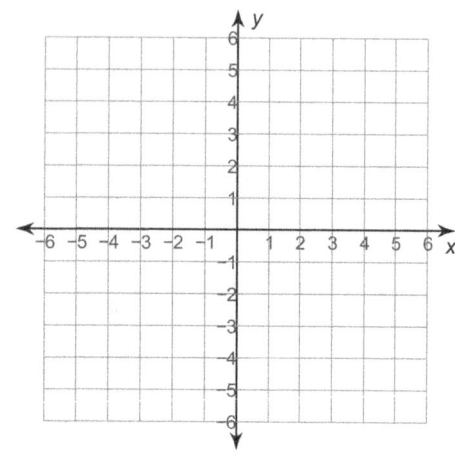

(888) $y \geq -\dfrac{3}{5}x + 1$

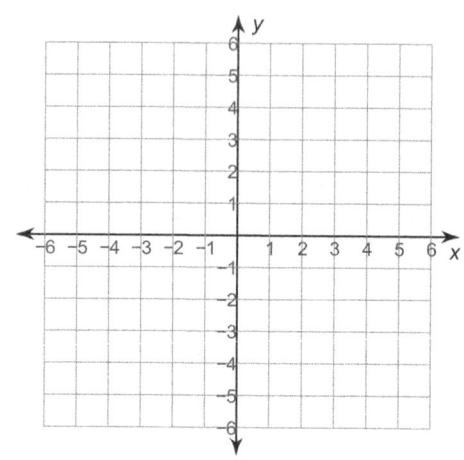

(889) $y \geq \dfrac{1}{2}x + 3$

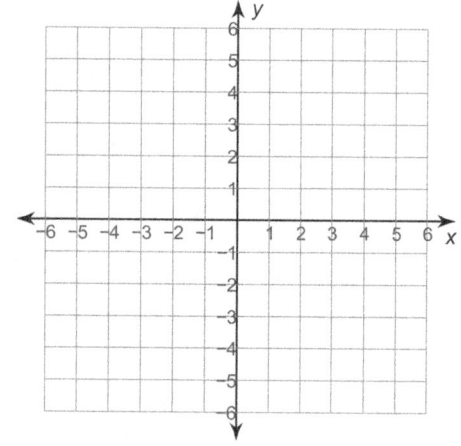

(890) $y < \dfrac{8}{3}x - 5$

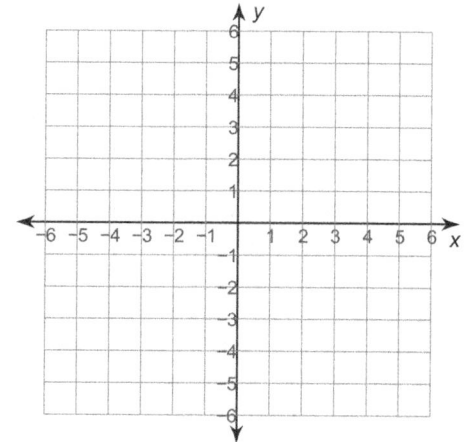

Graph each of the below absolute value equations.

(891) $y = |x+2|$

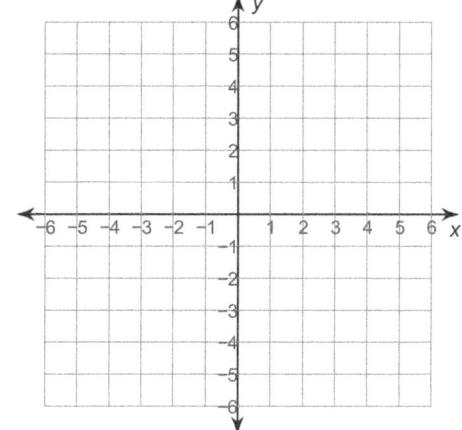

(892) $y = |x-4| + 3$

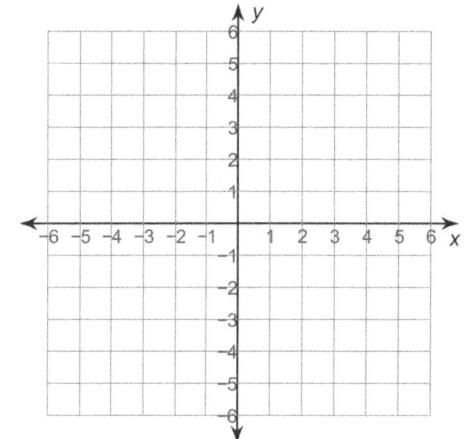

(893) $y = |x+1|$

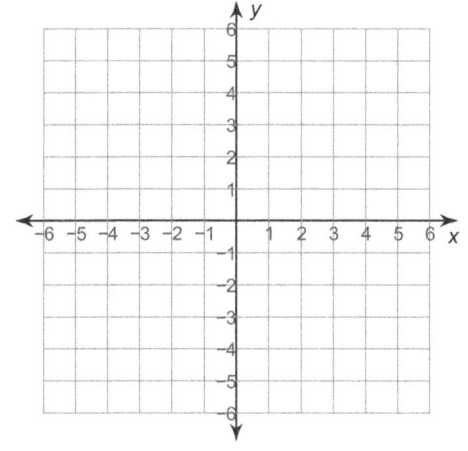

(894) $y = |x| + 1$

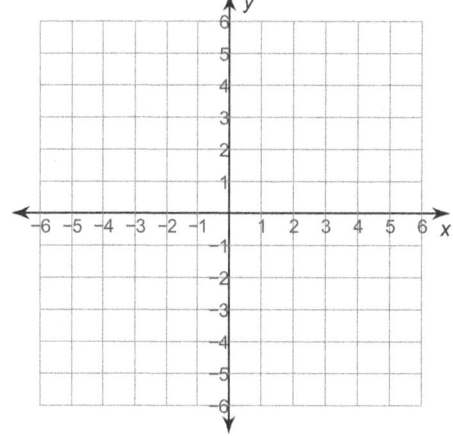

Graph each of the below absolute value equations.

(895) $y = |x - 3| + 3$

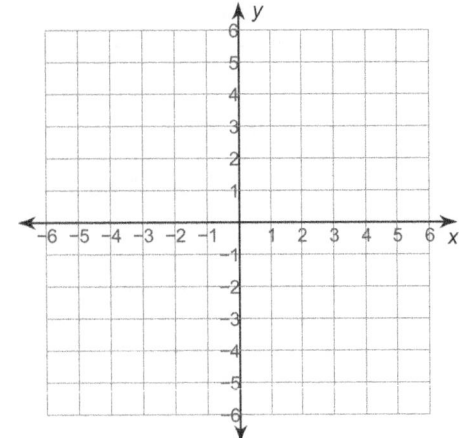

(896) $y = |x| + 4$

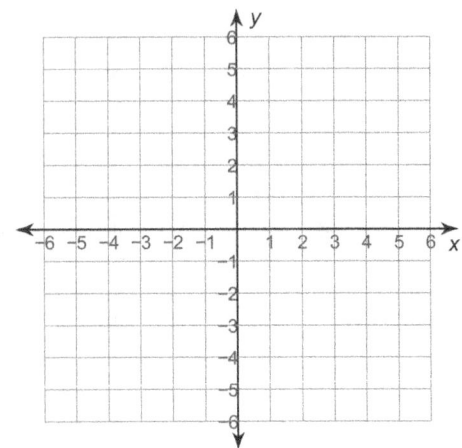

(897) $y = |x| - 2$

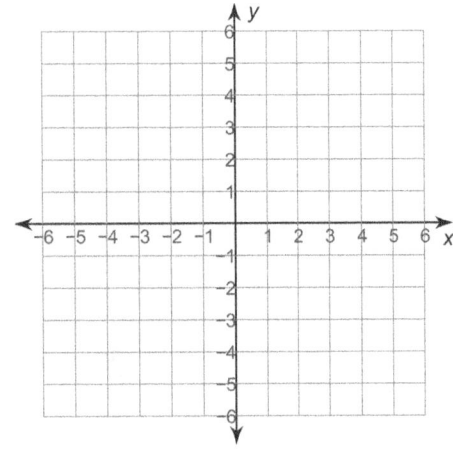

(898) $y = |x - 1| + 2$

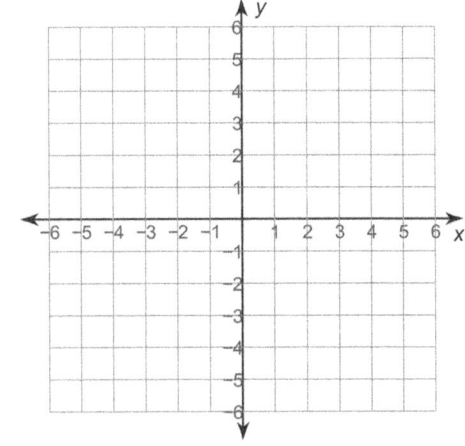

Graph each of the below absolute value equations.

(899) $y = |x - 2|$

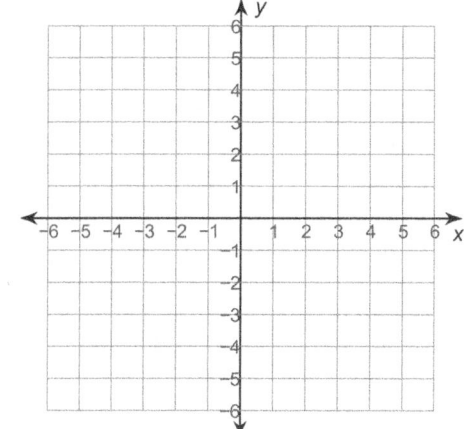

(900) $y = |x - 2| + 1$

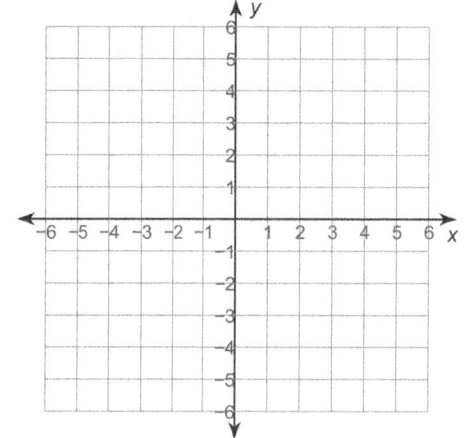

(901) $y = |x + 3|$

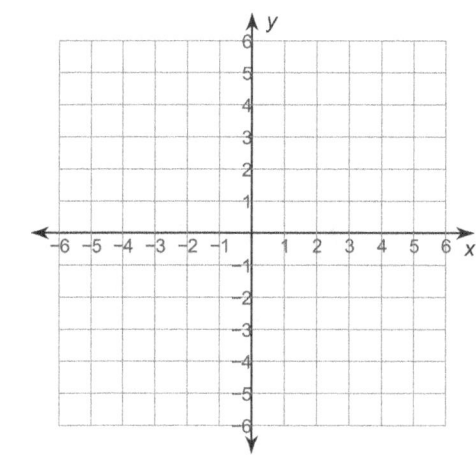

(902) $y = |x + 2| - 4$

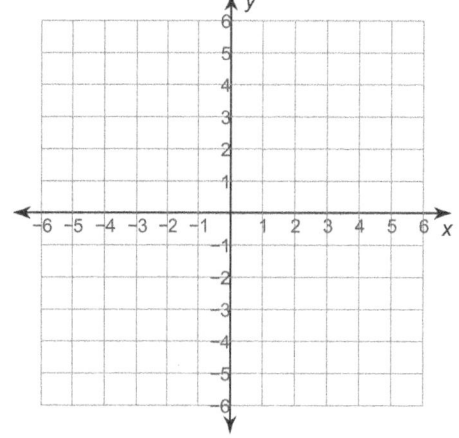

Graph each of the below absolute value equations.

(903) $y = |x| - 1$

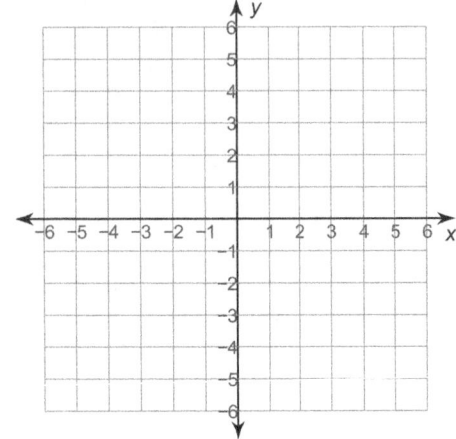

(904) $y = |x - 2| + 4$

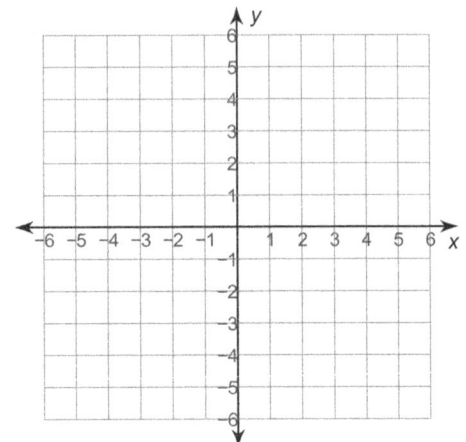

(905) $y = |x + 1| - 3$

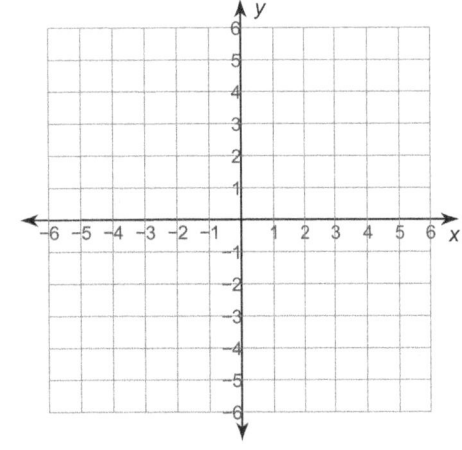

(906) $y = |x + 4| + 2$

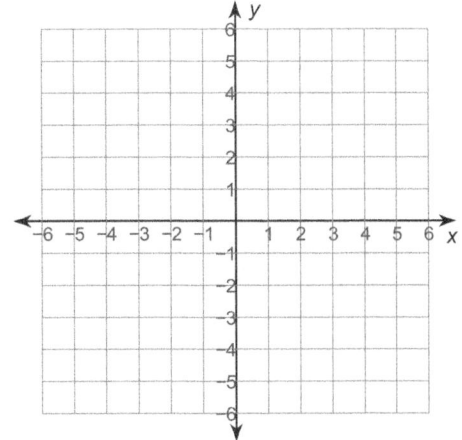

Graph each of the below absolute value equations.

(907) $y = |x + 3| - 2$

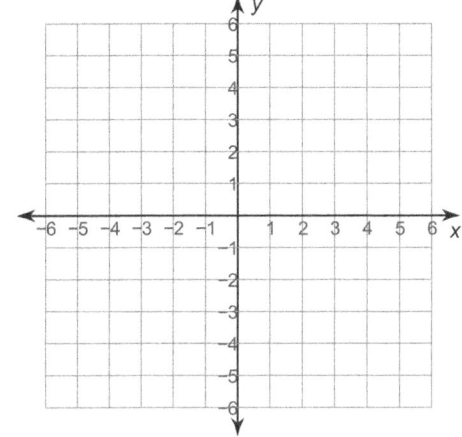

(908) $y = |x - 4| + 1$

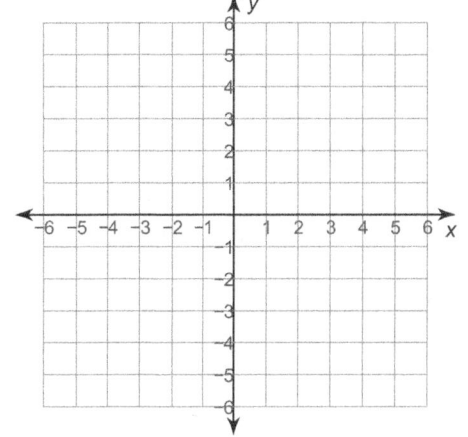

(909) $y = |x + 1| - 1$

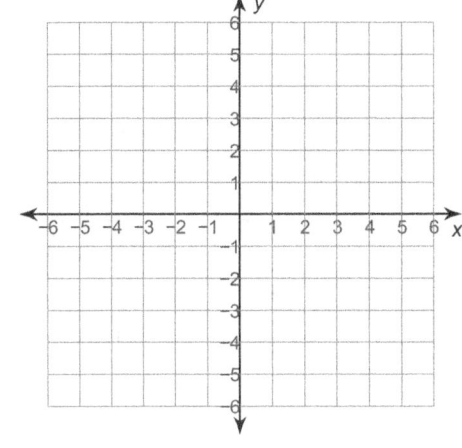

(910) $y = |x - 1|$

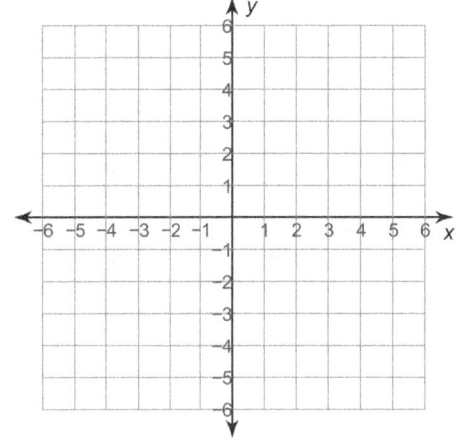

Graph each of the below absolute value equations.

(911) $y = |x - 4|$

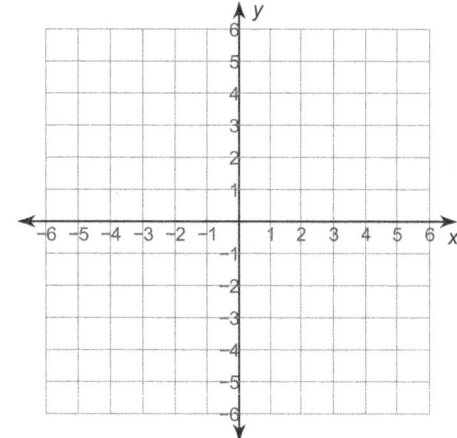

(912) $y = |x + 2| + 1$

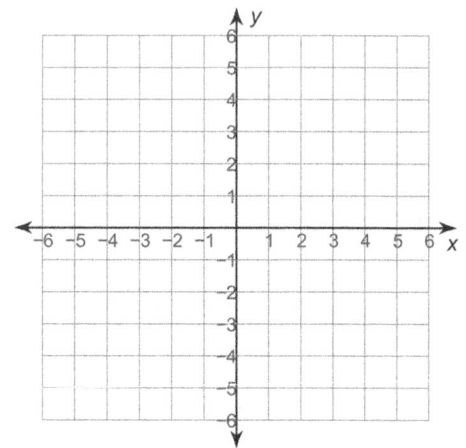

(913) $y = |x - 2| - 2$

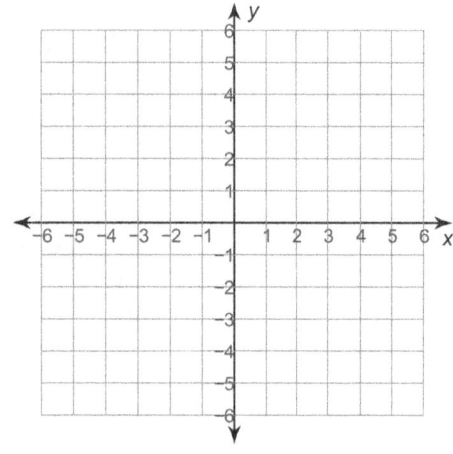

(914) $y = |x + 3| + 3$

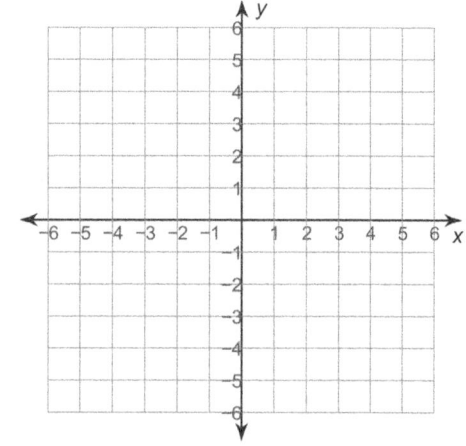

Graph each of the below absolute value equations.

(915) $y = |x| + 2$

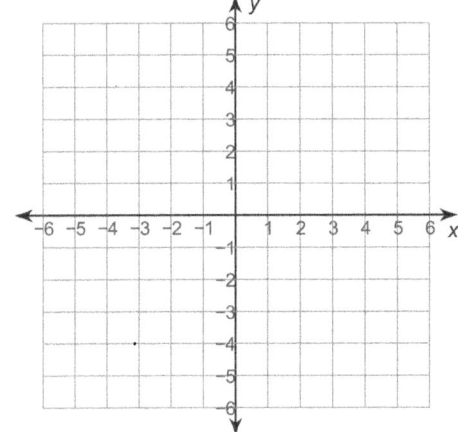

(916) $y = -|x| + 4$

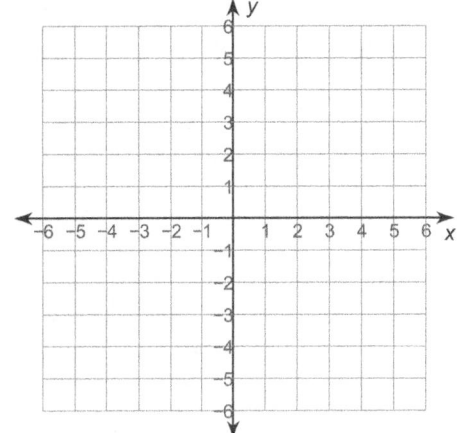

(917) $y = -|x| + 1$

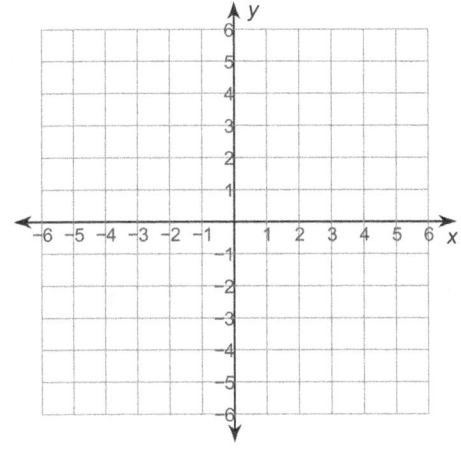

(918) $y = -|x - 2|$

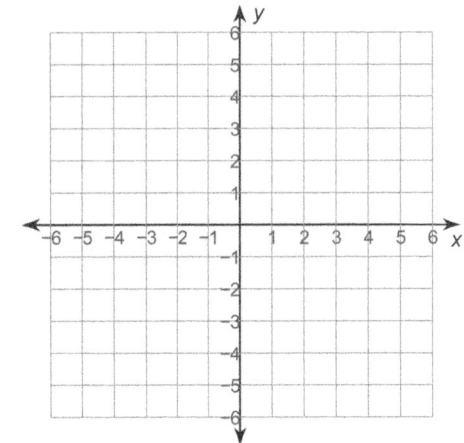

Graph each of the below absolute value equations.

(919) $y = -|x - 1| + 3$

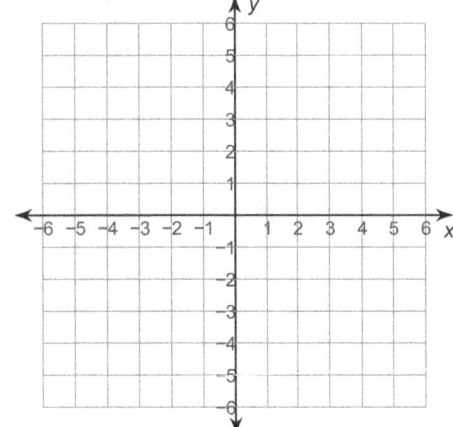

(920) $y = -|x - 4|$

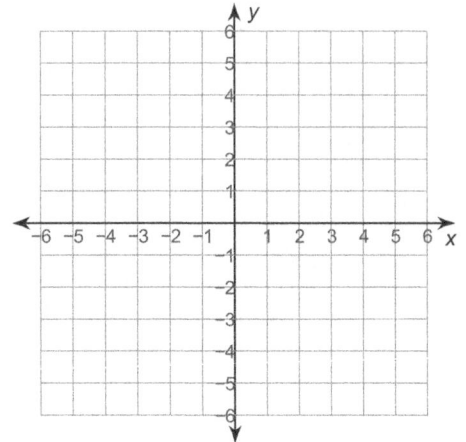

(921) $y = -|x + 3| + 4$

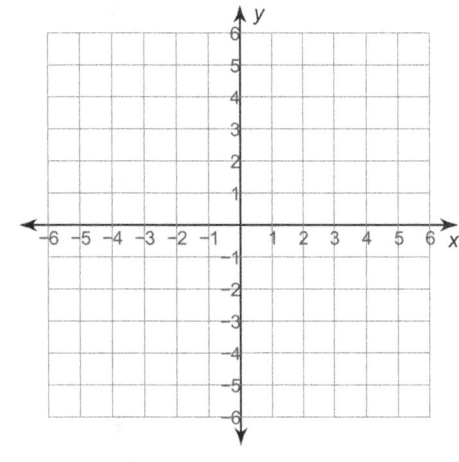

(922) $y = -|x - 2| - 2$

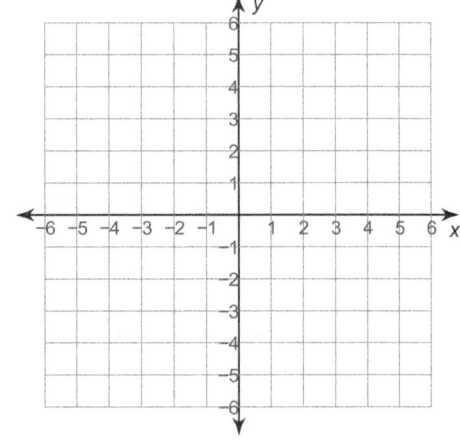

Graph each of the below absolute value equations.

(923) $y = -|x-3|$

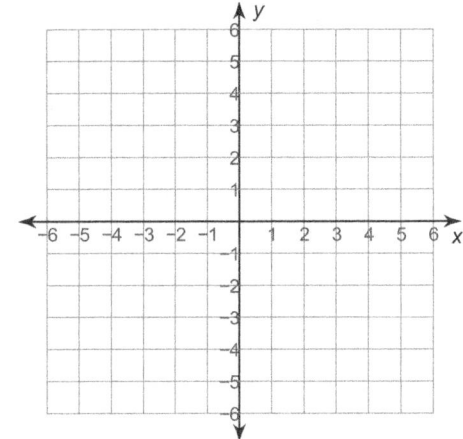

(924) $y = -|x-3| - 4$

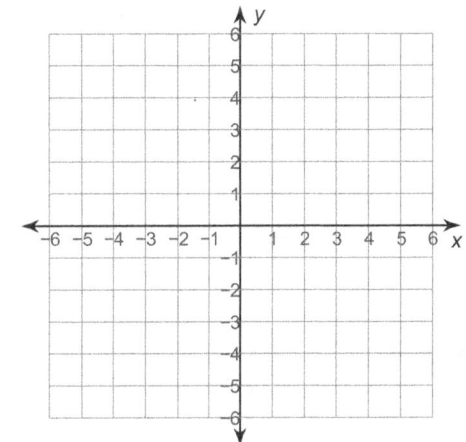

(925) $y = -|x+1| - 3$

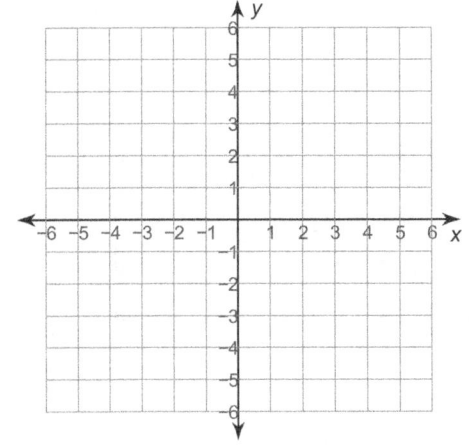

(926) $y = -|x+4|$

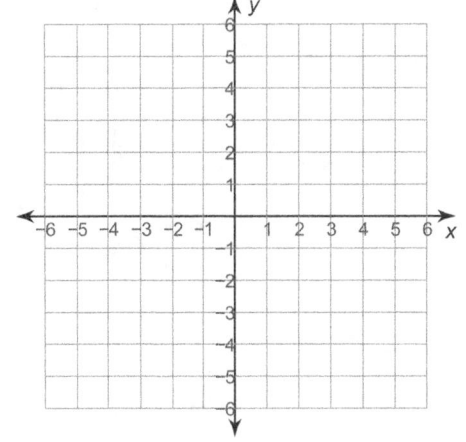

Graph each of the below absolute value equations.

(927) $y = -|x + 1|$

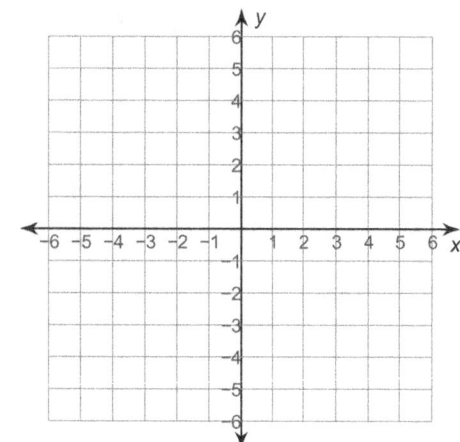

(928) $y = -|x| + 2$

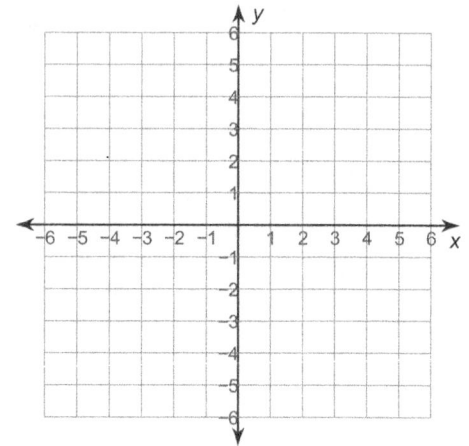

(929) $y = -|x + 2|$

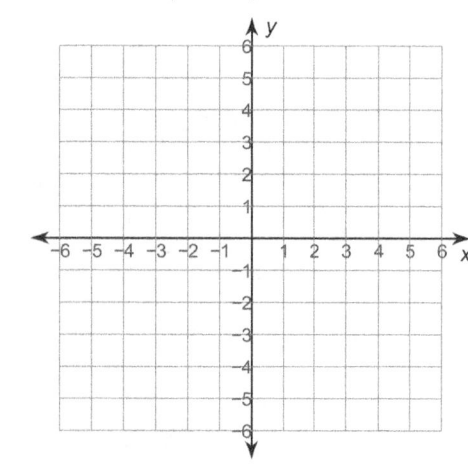

(930) $y = -|x + 4| + 1$

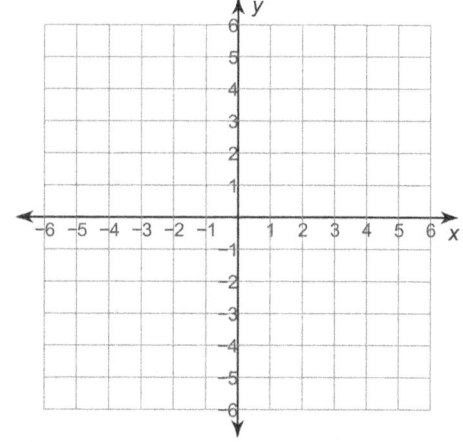

Graph each of the below absolute value equations.

(931) $y = -|x - 2| - 3$

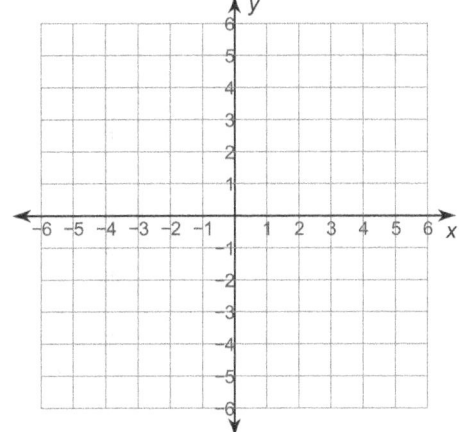

(932) $y = -|x + 2| + 2$

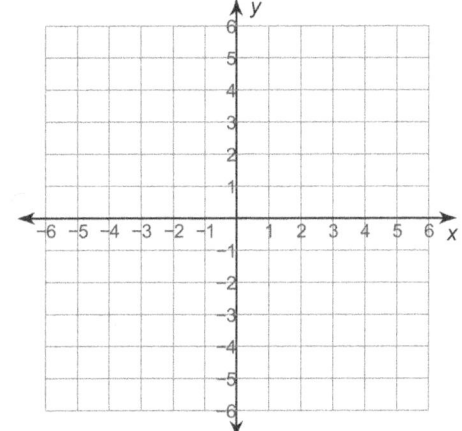

(933) $y = -|x - 3| + 3$

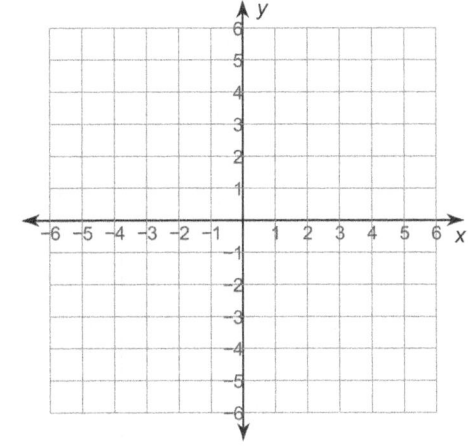

(934) $y = -|x| - 2$

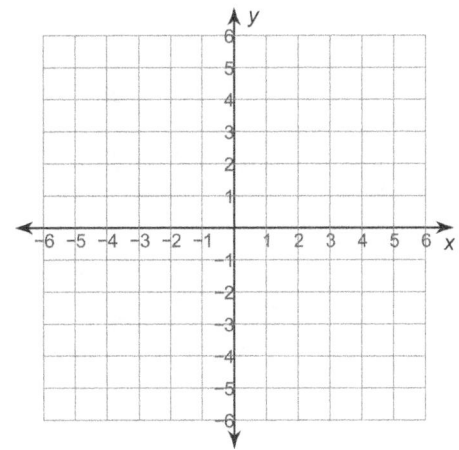

Graph each of the below absolute value equations.

(935) $y = -|x| - 4$

(936) $y = -|x| + 3$

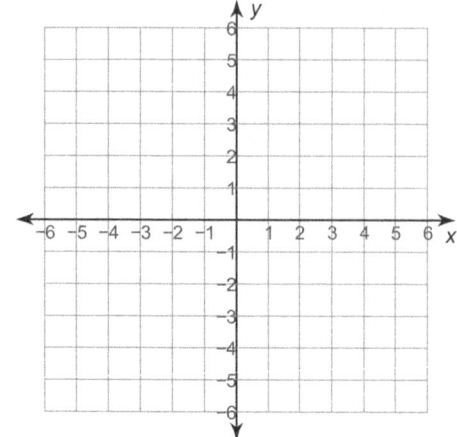
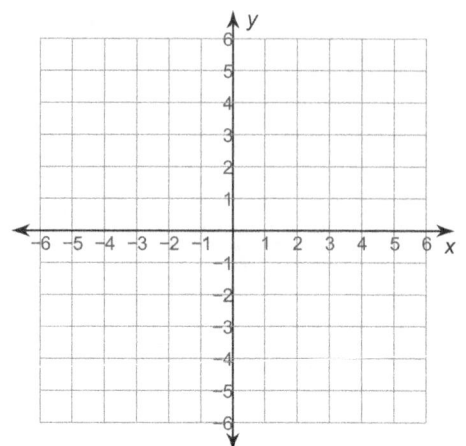

(937) $y = -|x - 4| + 1$

(938) $y = -|x - 2| + 3$

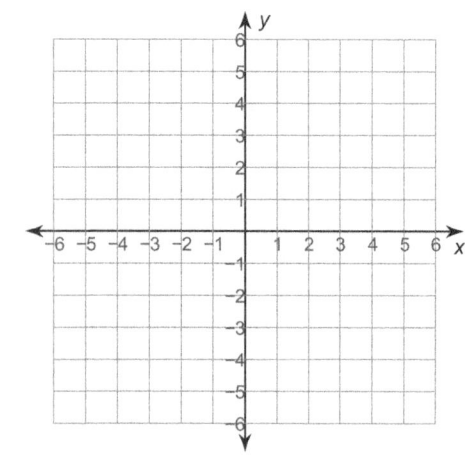
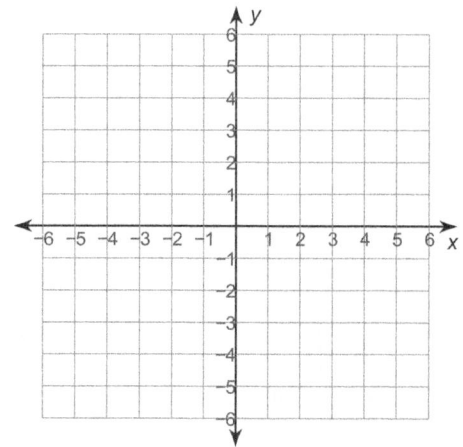

Graph each of the below absolute value equations.

(939) $y = -|x + 3| - 2$

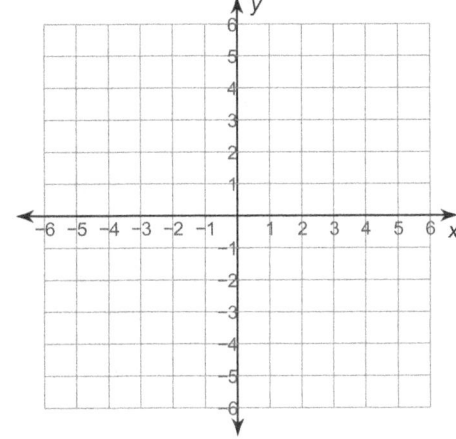

(940) $y = -|x + 3| - 4$

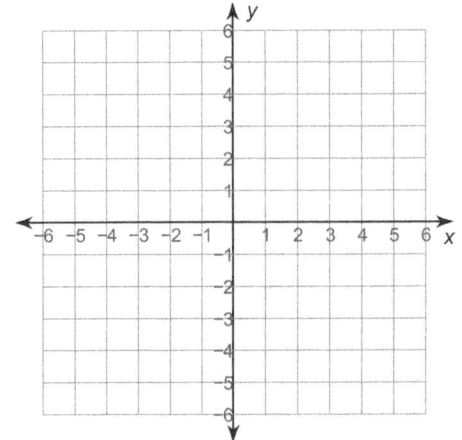

Solve the below system of equations by graphing.

(941) $x - 2y = 2$
$7x - 4y = -16$

(942) $x - 2y = -2$
$3x - 2y = 6$

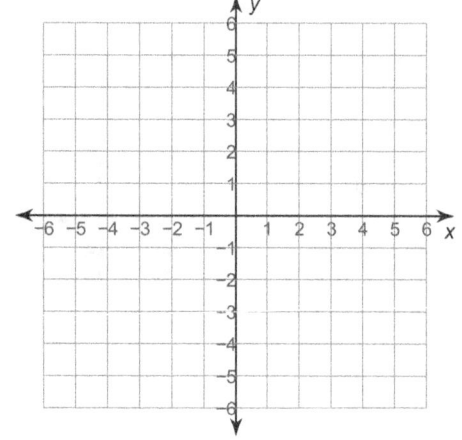
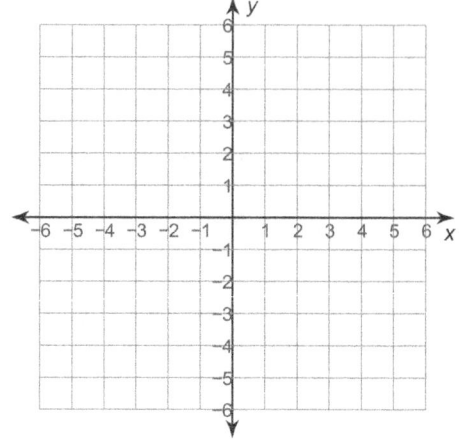

(943) $4x + y = 2$
$y = -2$

(944) $2x - y = -1$
$2x - y = 2$

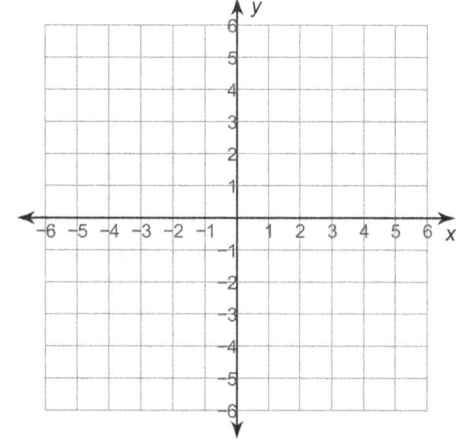
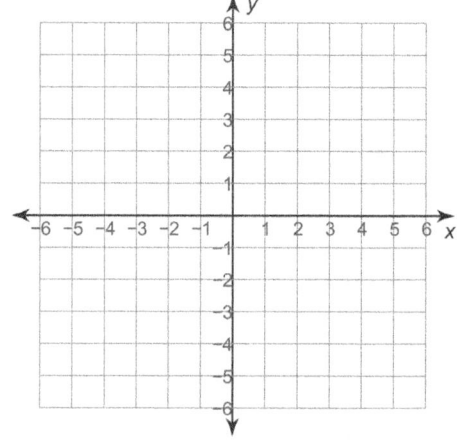

Solve the below system of equations by graphing.

(945) $x + 2y = -6$
       $3x + 2y = -2$

(946) $7x + y = -3$
       $x + y = 3$

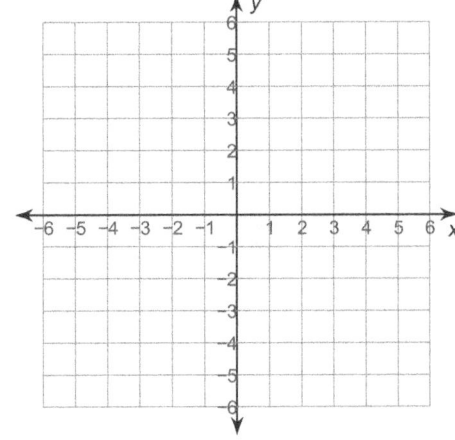
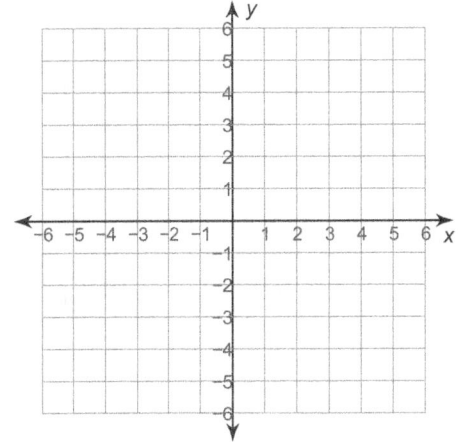

(947) $7x + 2y = -6$
       $x + 2y = 6$

(948) $7x - 2y = -6$
       $x - 2y = 6$

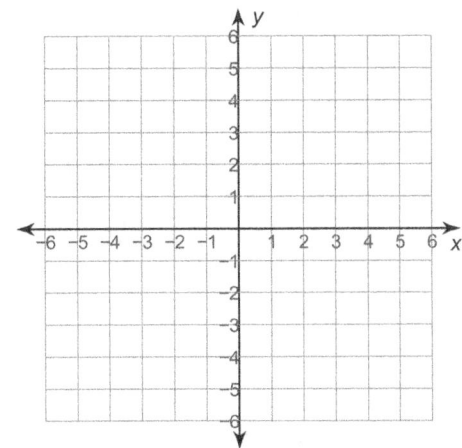
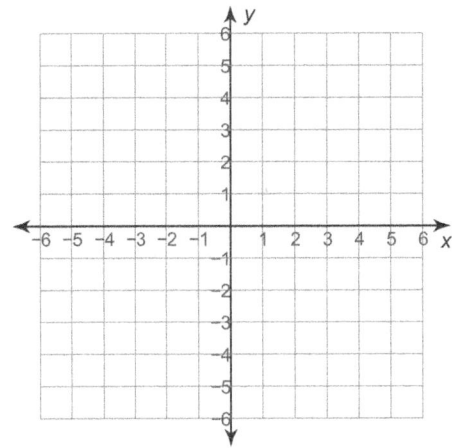

Solve the below system of equations by graphing.

(949) $2x - y = 2$
       $x + 2y = 6$

(950) $7x + 3y = -9$
       $2x + 3y = 6$

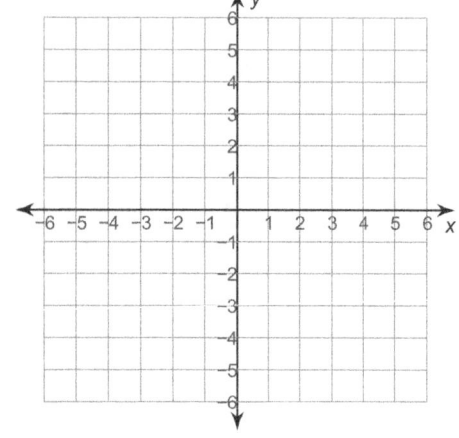
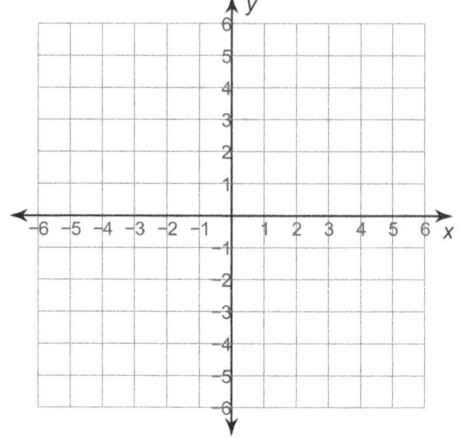

(951) $5x - 2y = -6$
       $x - 2y = 2$

(952) $2x - y = -3$
       $5x + y = -4$

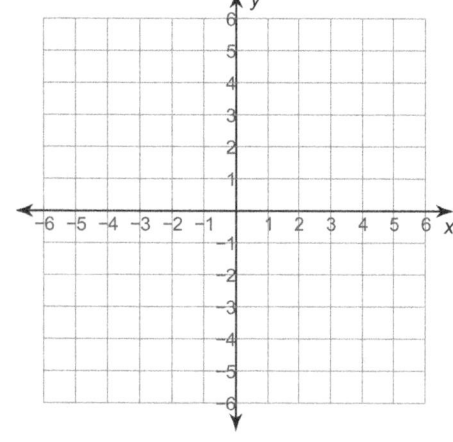
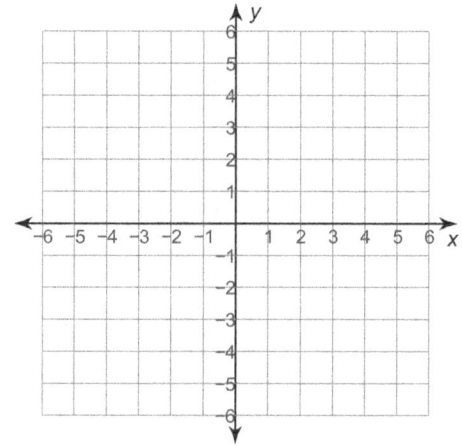

Solve the below system of equations by graphing.

(953) $5x + 2y = 2$
       $x + 2y = -6$

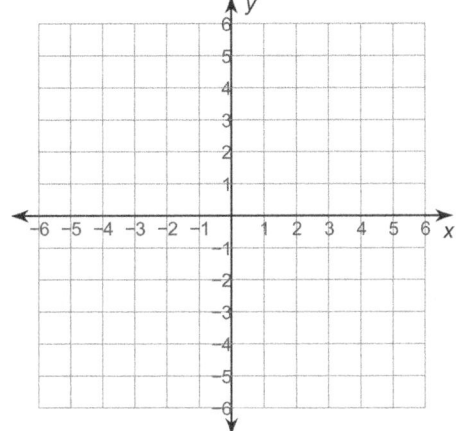

(954) $x - y = -4$
       $2x + 3y = -3$

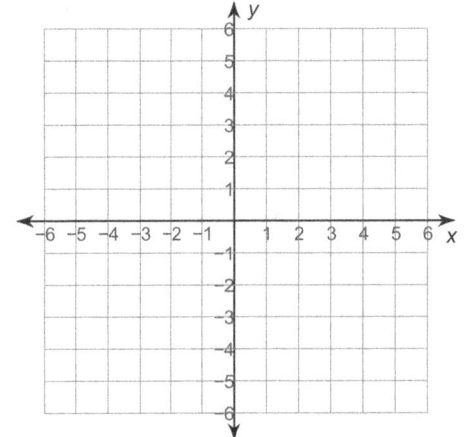

(955) $2x - y = -1$
       $2x - y = 4$

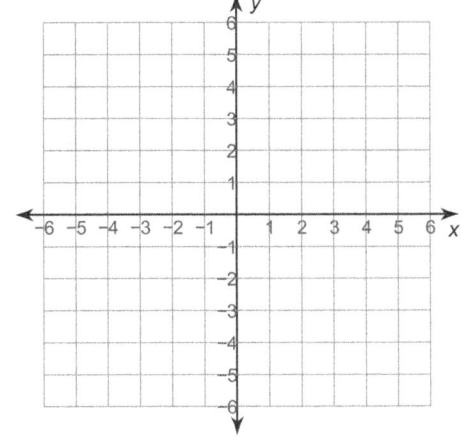

(956) $x + y = -3$
       $x + y = 2$

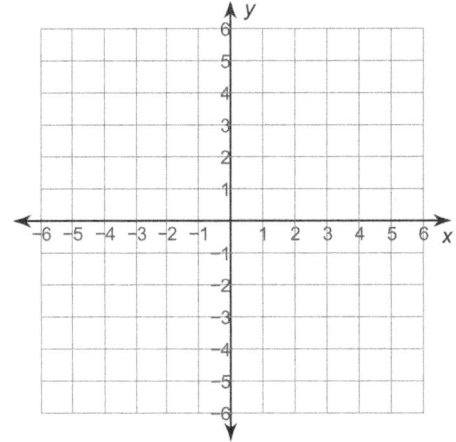

Solve the below system of equations by graphing.

(957) $4x + 3y = -9$
       $2x - 3y = -9$

(958) $7x - 4y = 12$
       $x - 4y = -12$

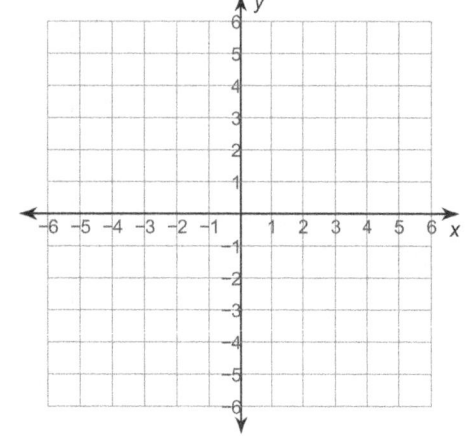
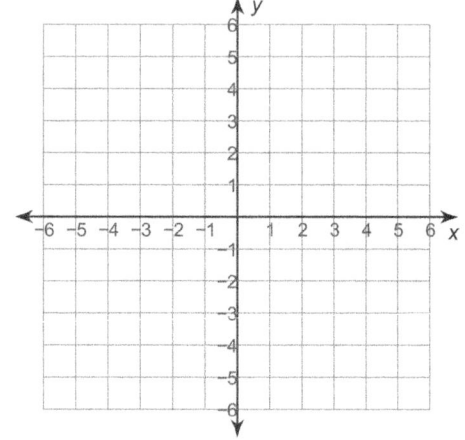

(959) $x + 2y = 6$
       $3x + 2y = 2$

(960) $5x - y = 3$
       $2x + y = 4$

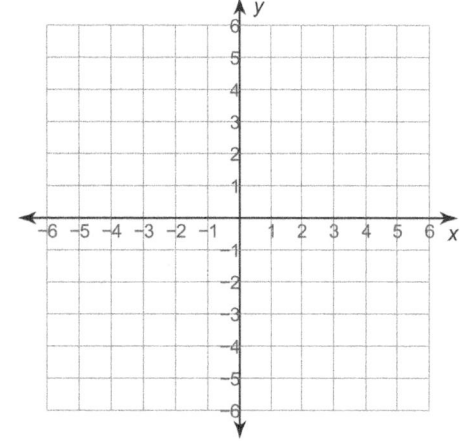
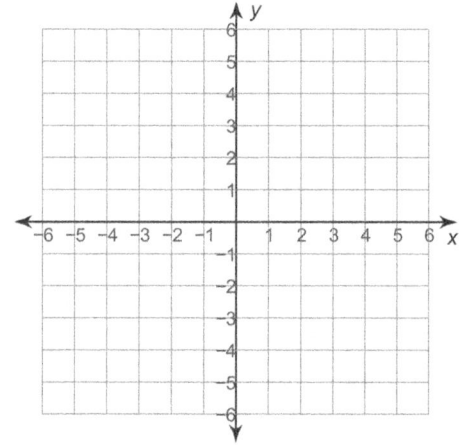

Solve the below system of equations by graphing.

(961) $7x + 3y = 12$
       $7x + 3y = 9$

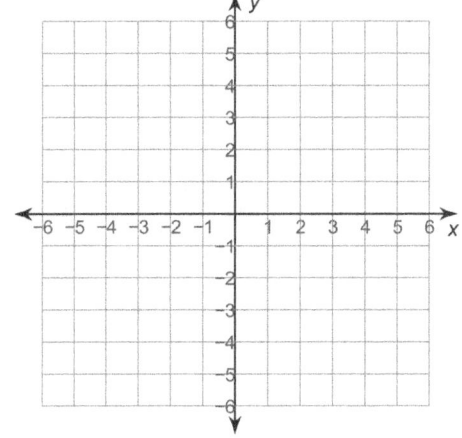

(962) $3x + 4y = -4$
       $x - 4y = -12$

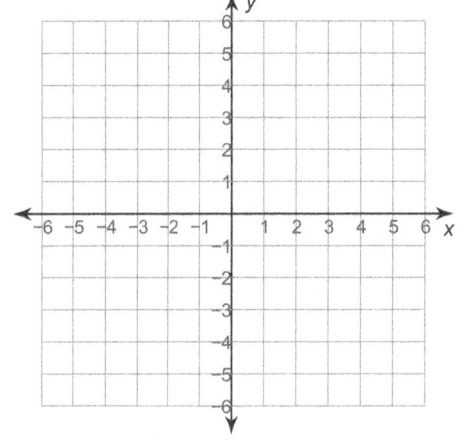

(963) $7x - y = 4$
       $x - y = -2$

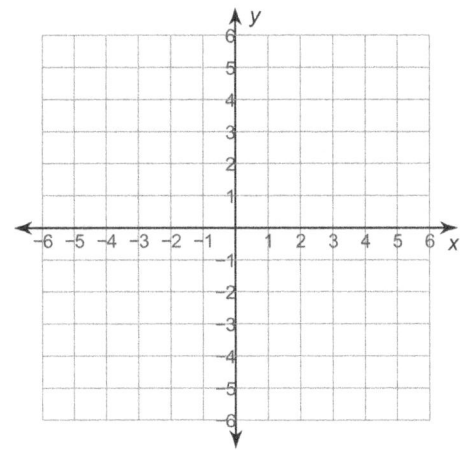

(964) $5x + y = 3$
       $x - y = 3$

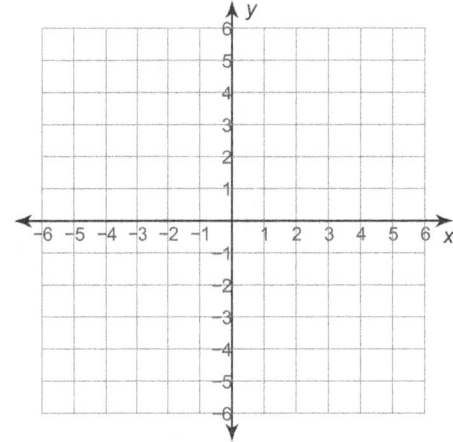

Solve the below system of equations by graphing.

(965) $x + 4y = 8$
       $x - 2y = 2$

(966) $2x + y = 1$
       $2x - y = 3$

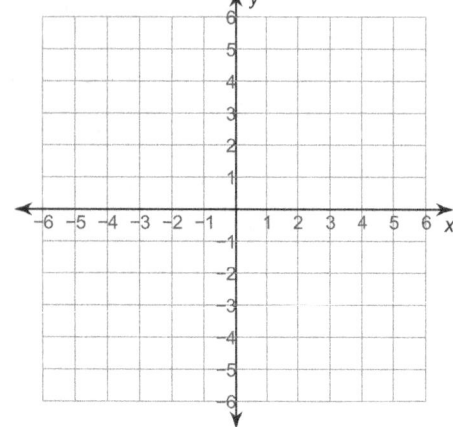
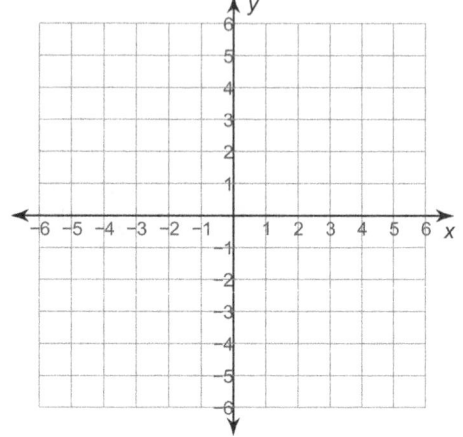

(967) $x + y = -4$
       $x + y = 4$

(968) $7x + y = 4$
       $x - y = 4$

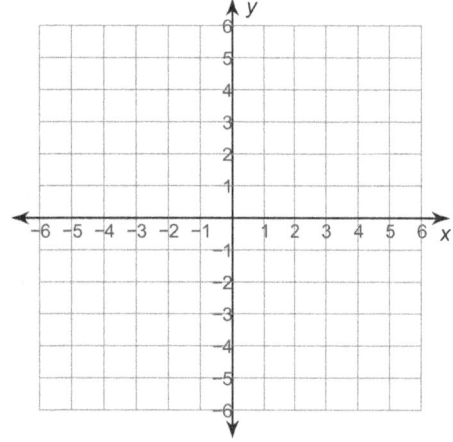
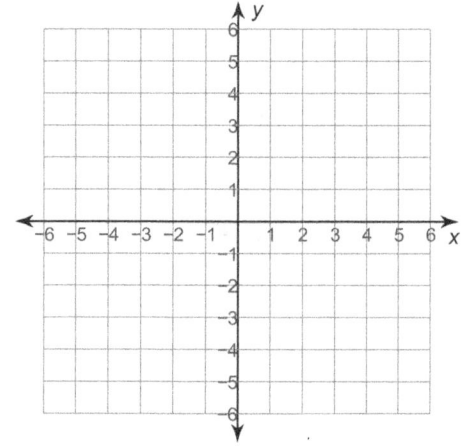

Solve the below system of equations by graphing.

(969) $3x + 4y = -4$
       $3x + 4y = -8$

(970) $x - 4y = 8$
       $x - y = -1$

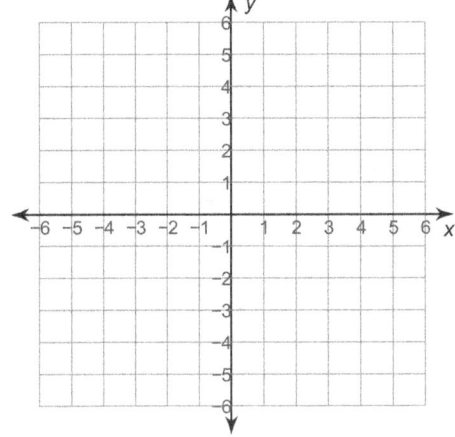
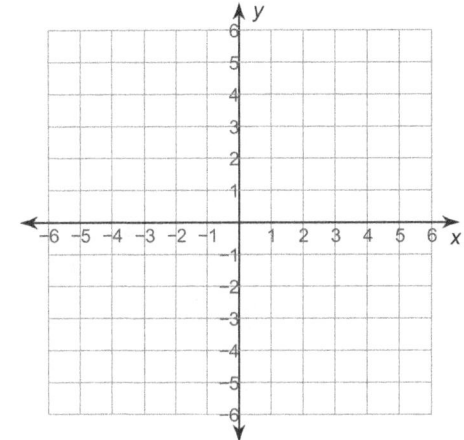

(971) $7x - 9y = -36$
       $x = -9$

(972) $2x + y = -11$
       $3x - 10y = -120$

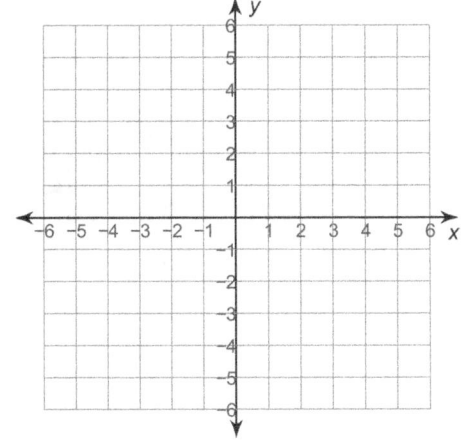
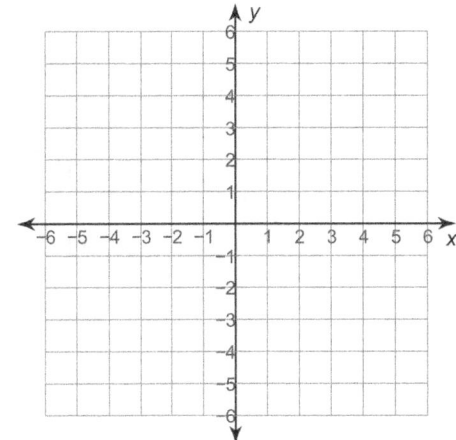

Solve the below system of equations by graphing.

(973) $25x + 11y = -77$
       $7x + 11y = 121$

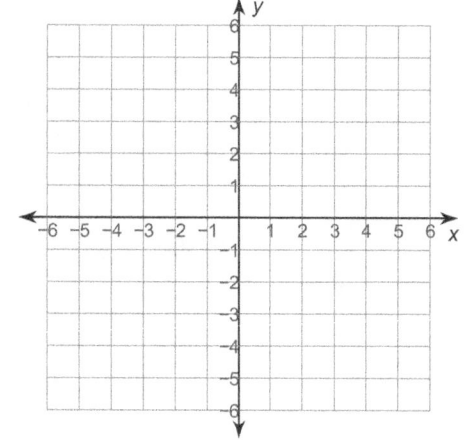

(974) $8x - 17y = 187$
       $21x + 17y = 306$

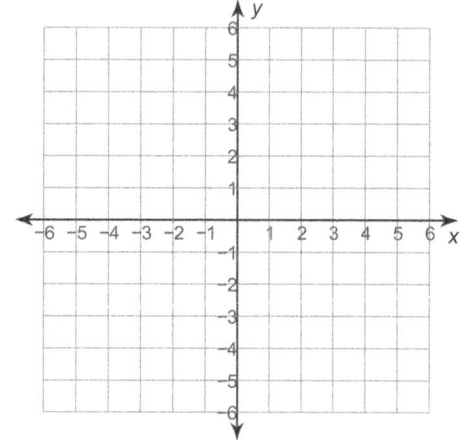

(975) $3x - 2y = 4$
       $x = -10$

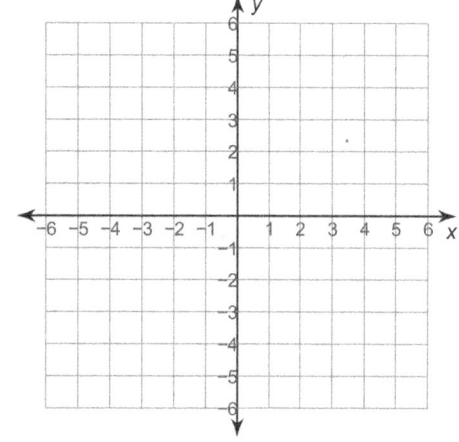

(976) $17x + 14y = 266$
       $8x - 7y = 98$

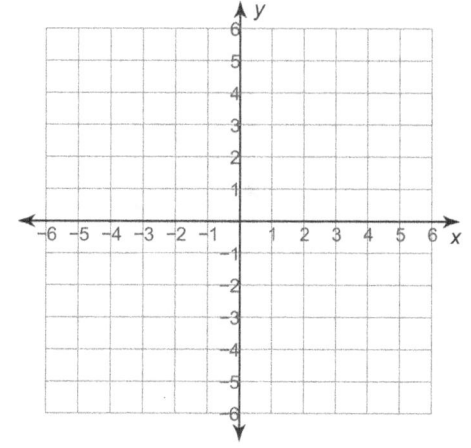

Solve the below system of equations by graphing.

(977) $22x - 9y = -162$
$x - 9y = 27$

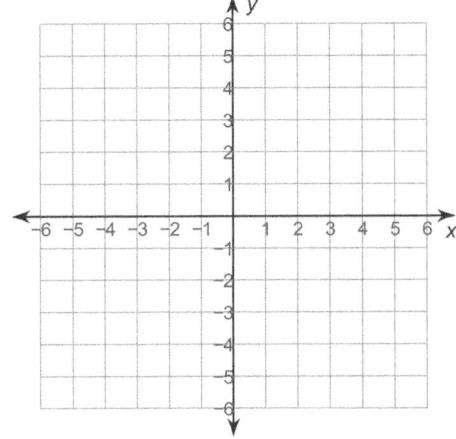

(978) $6x - 7y = 70$
$11x + 14y = 182$

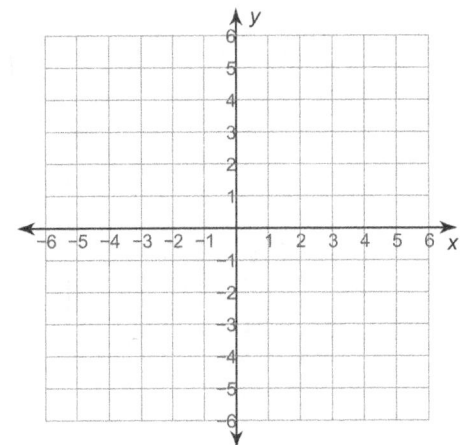

(979) $4x + 3y = 48$
$31x - 3y = 57$

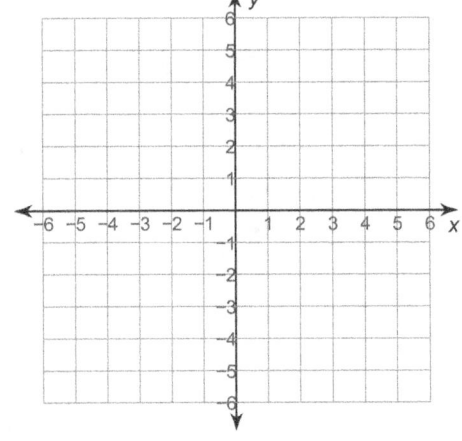

(980) $3x + y = -17$
$13x - 5y = -55$

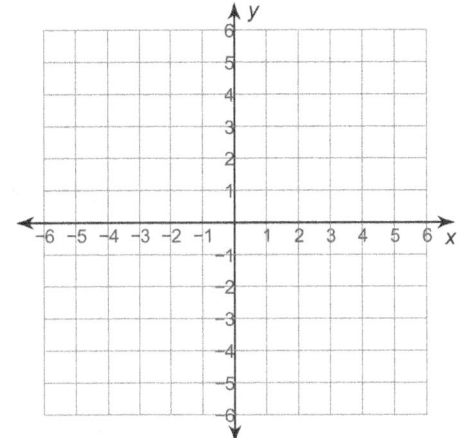

Solve the below system of equations by graphing.

(981) $35x + 12y = 204$
        $35x + 12y = -36$

(982) $5x + 3y = 27$
        $5x + 3y = 42$

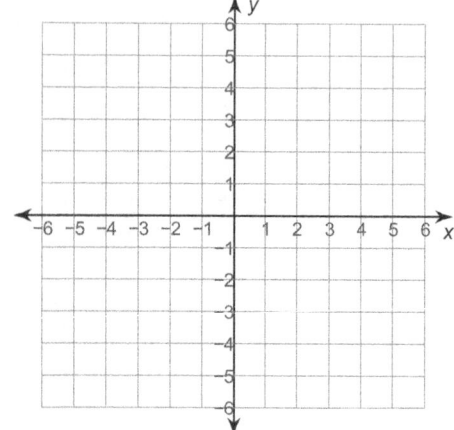
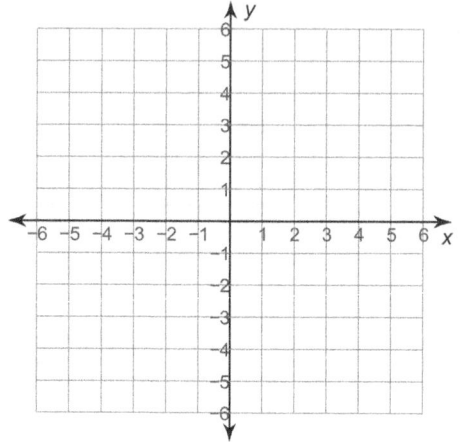

(983) $x - 14y = -196$
        $10x - 7y = 35$

(984) $x + 9y = 117$
        $7x - 3y = 27$

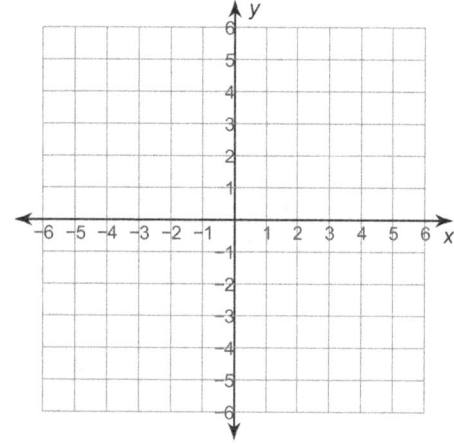
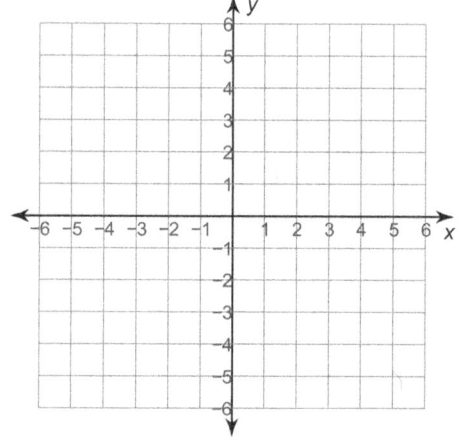

Solve the below system of equations by graphing.

(985) $24x - 19y = 285$
       $2x + 19y = 209$

(986) $x - 4y = -72$
       $17x - 2y = 30$

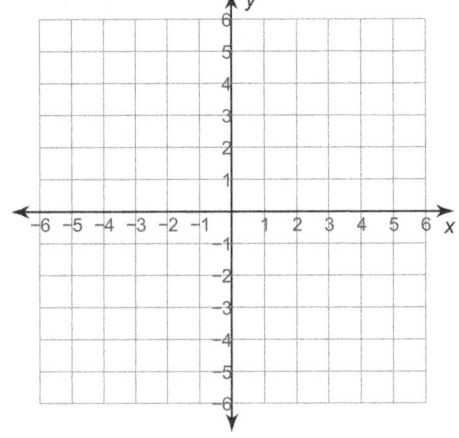
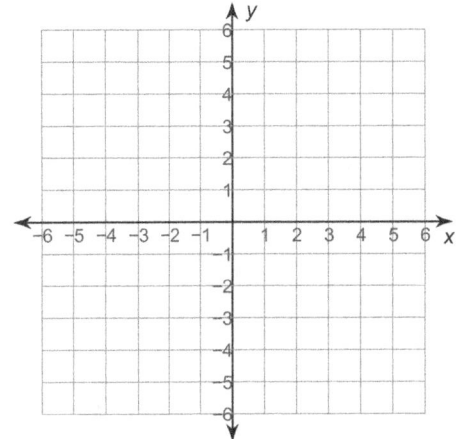

(987) $x + 3y = -27$
       $3x + y = 15$

(988) $11x + 3y = -15$
       $x + 3y = -45$

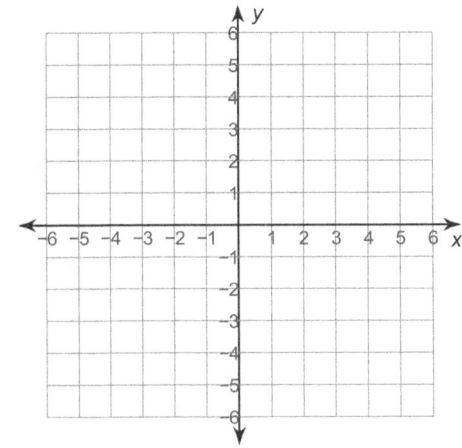
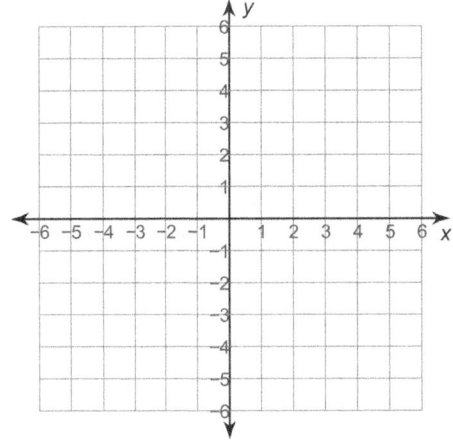

Solve the below system of equations by graphing.

(989) $x + 14y = 140$
       $5x + 7y = 133$

(990) $x - 7y = 70$
       $19x - 14y = -98$

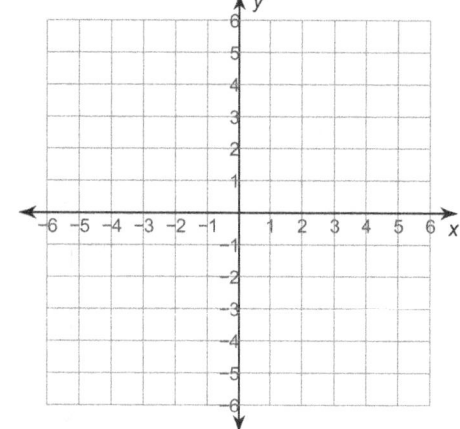
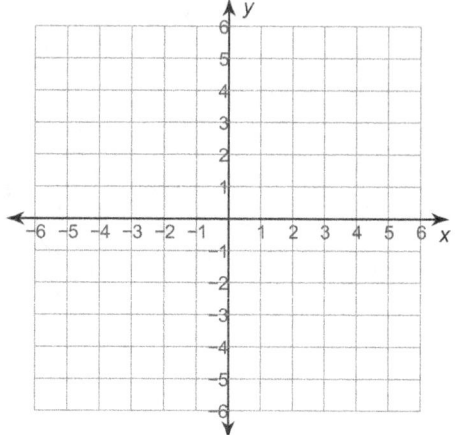

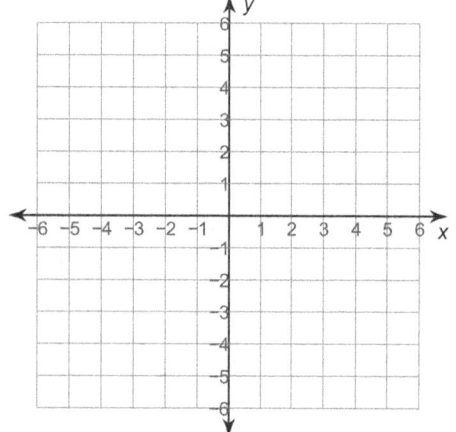
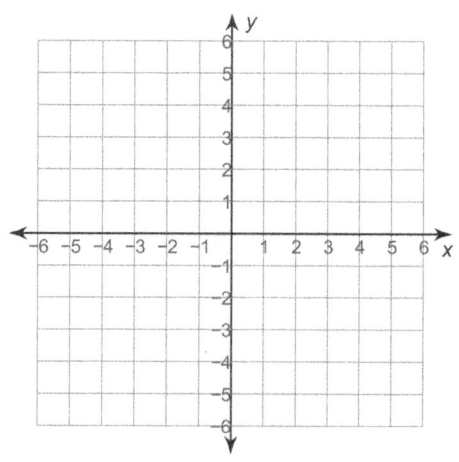

Solve the system of equations given below by using method of elimination.

(991) $12x - 2y = -20$
       $15x + 14y = 41$

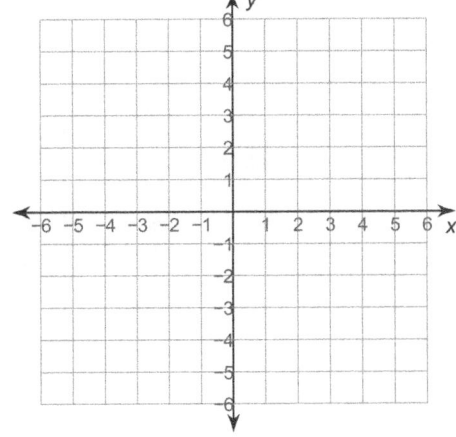

(992) $6x - 5y = -19$
       $-12x + 10y = 12$

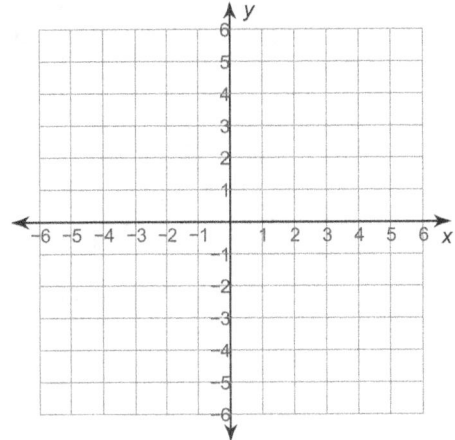

(993) $5x - 6y = -34$
       $-14x + 12y = 28$

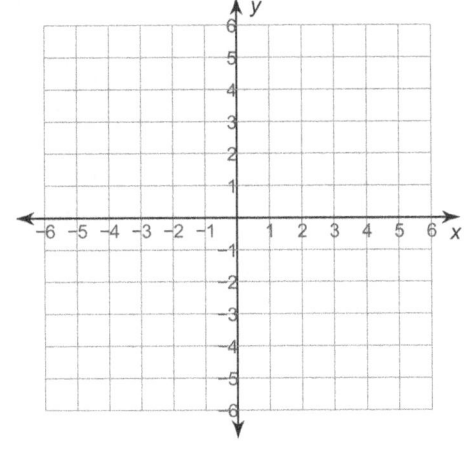

(994) $15x + 10y = 15$
       $x - 30y = 1$

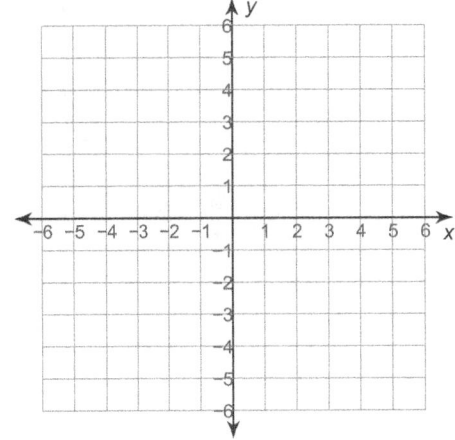

Solve the system of equations given below by using method of elimination.

(995) $-18x + 5y = 26$
       $-9x - 8y = 34$

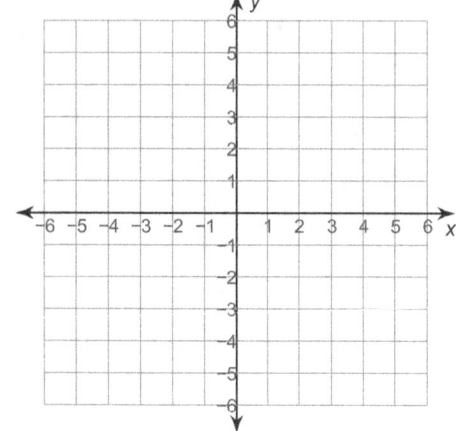

(996) $5x - 5y = 24$
       $15x - 15y = 45$

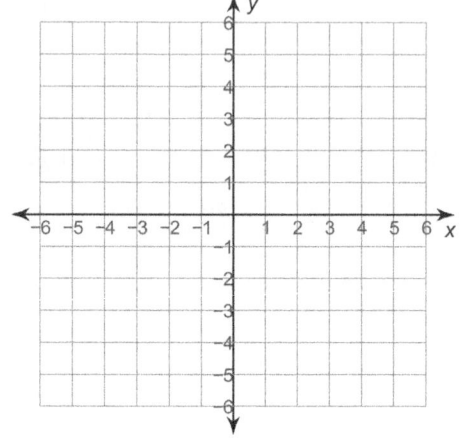

(997) $5x + 13y = 15$
       $8x + 26y = 24$

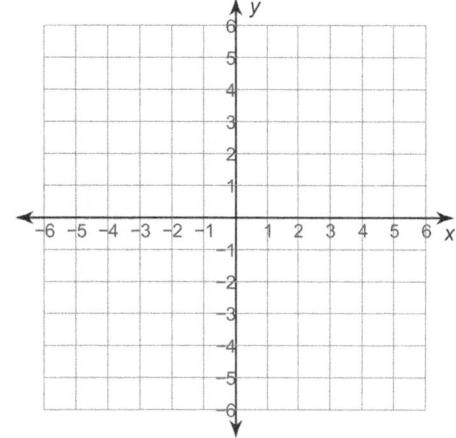

(998) $14x + 6y = 38$
       $-2x + 2y = -14$

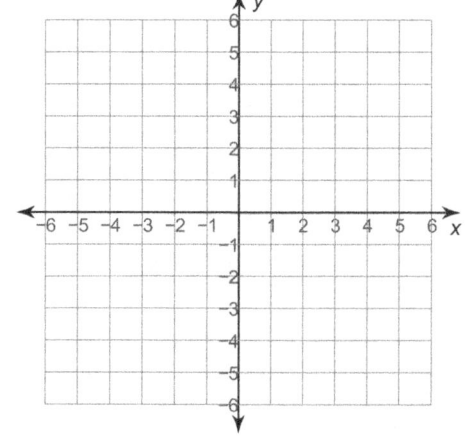

Solve the system of equations given below by using method of elimination.

(999) $-11x - 38y = -21$
       $-16x - 19y = 42$

(1000) $-11x + 4y = -50$
       $-18x + 12y = 0$

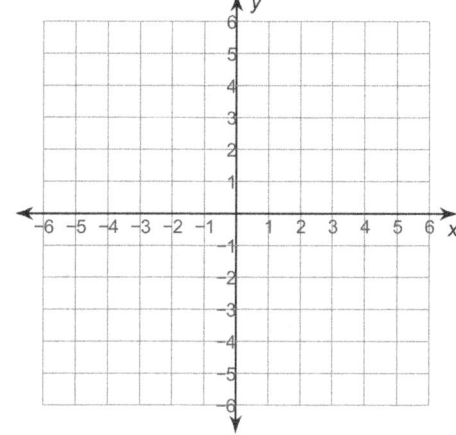
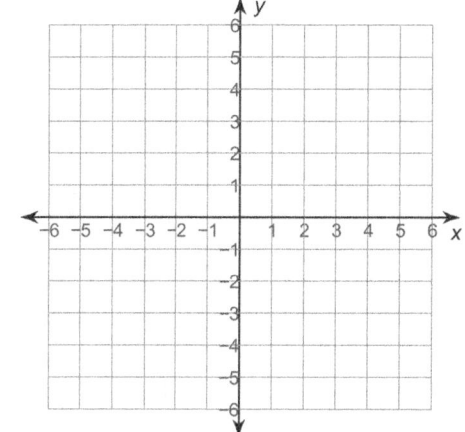

(1001) $21x - 15y = -57$
       $7x - 8y = -43$

(1002) $-x - 8y = 17$
       $6x + 16y = 26$

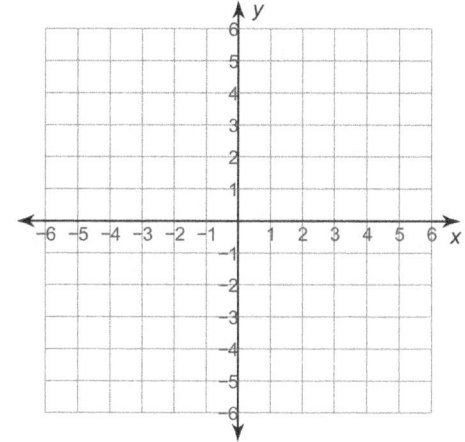
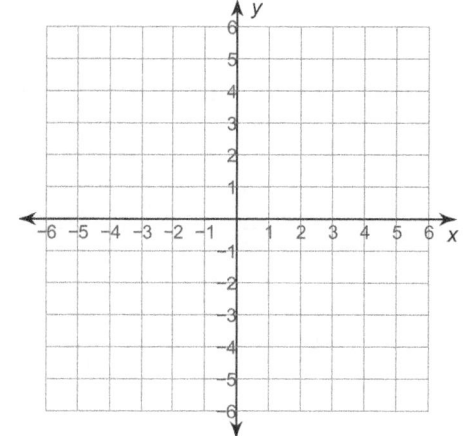

Solve the system of equations given below by using method of elimination.

(1003) $x - 6y = -24$
         $6x + 18y = 18$

(1004) $-7x + 9y = -1$
         $13x - 27y = -29$

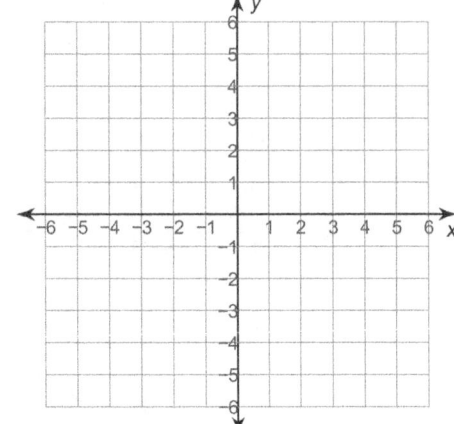
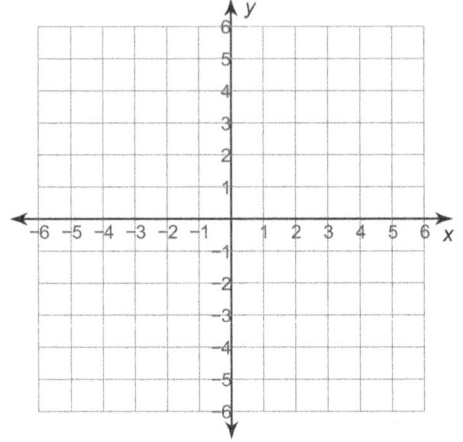

(1005) $-6x - 18y = -31$
         $-3x - 9y = -21$

(1006) $4x - 17y = 13$
         $-16x + 34y = -18$

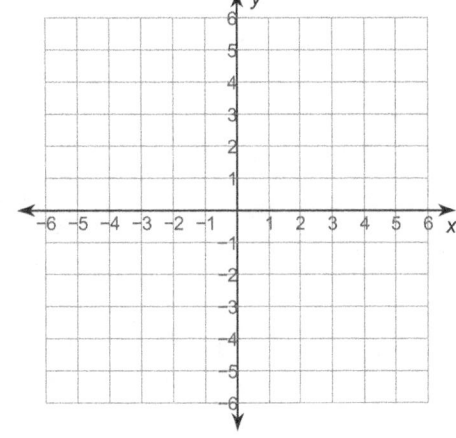
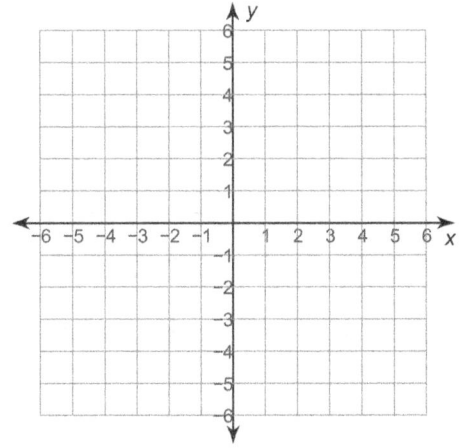

Solve the system of equations given below by using method of elimination.

(1007) $-18x + 4y = 2$
         $-20x + 2y = -10$

(1008) $-11x + 24y = 5$
         $3x + 12y = -57$

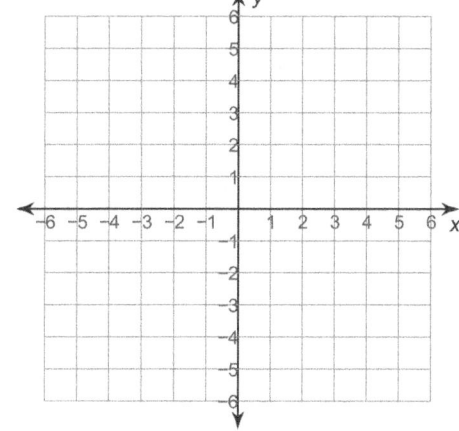
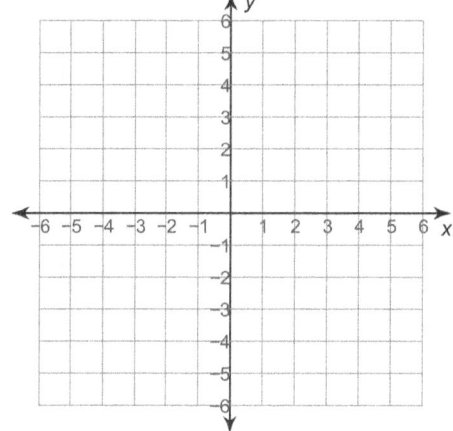

(1009) $-5x - 8y = -26$
         $-15x + 5y = -20$

(1010) $2x + 9y = -45$
         $20x + 18y = 54$

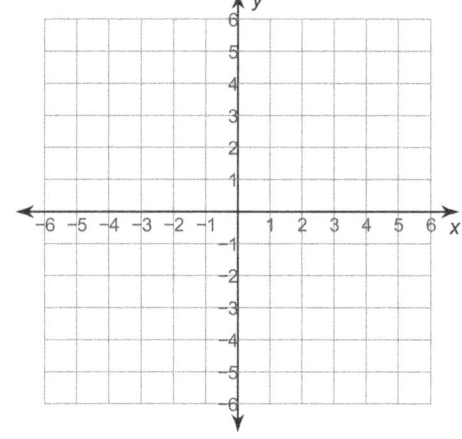
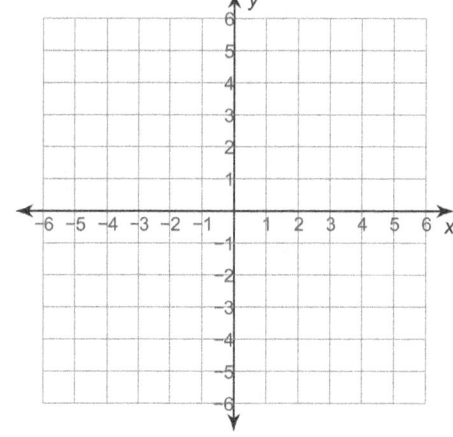

Solve the system of equations given below by using method of elimination.

(1011) $-7x + 11y = -6$
         $-2x + 17y = 26$

(1012) $-3x + 5y = 4$
         $20x + 6y = 52$

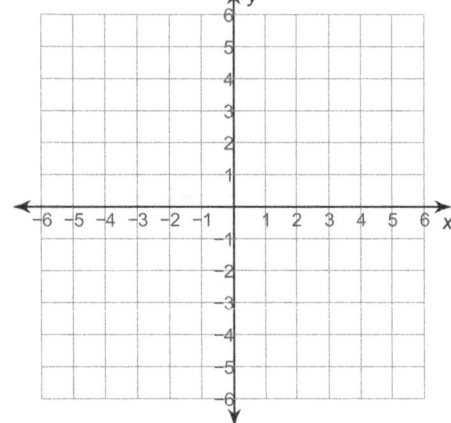
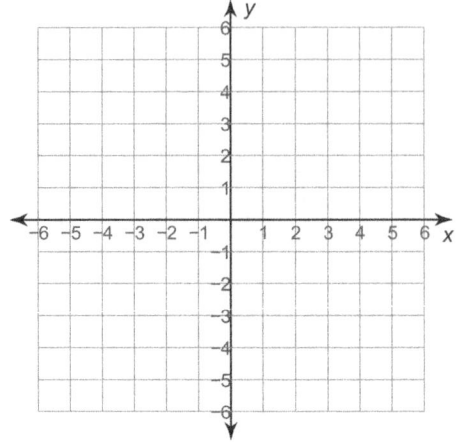

(1013) $20x - 4y = -56$
         $-14x - 15y = -32$

(1014) $-10x + 3y = -7$
         $-15x + 16y = 1$

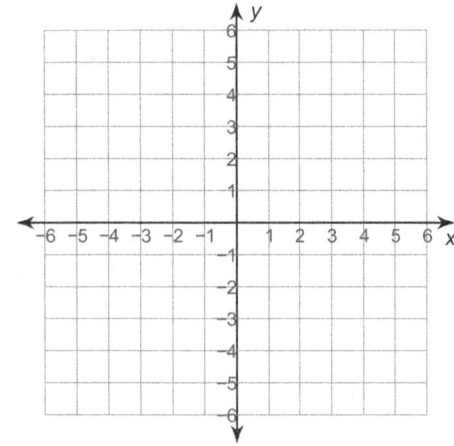
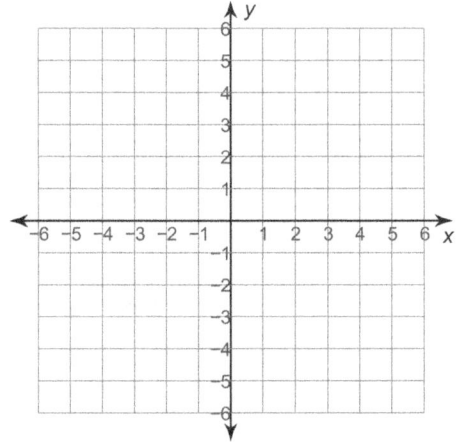

Solve the system of equations given below by using method of elimination.

(1015) $9x - 2y = 42$
$-7x + 9y = 12$

(1016) $3x - 5y = -31$
$5x - 8y = -46$

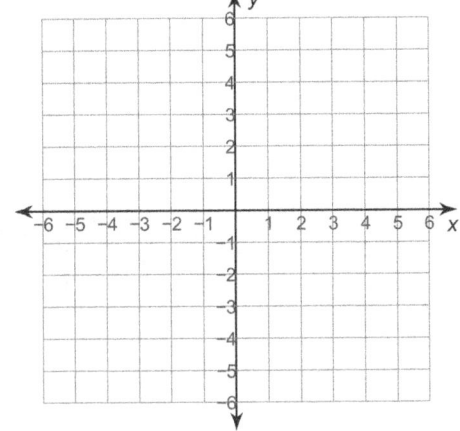
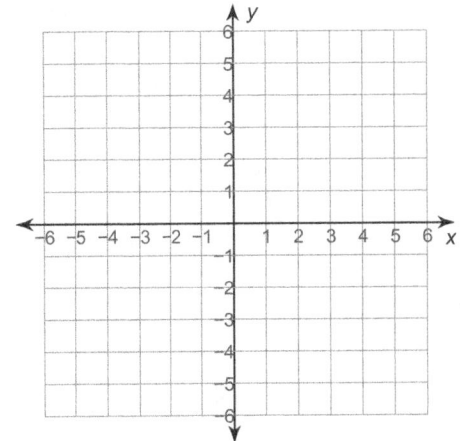

(1017) $-15x + 4y = 53$
$-2x - 17y = -28$

(1018) $-5x - 7y = 43$
$-12x + 20y = -44$

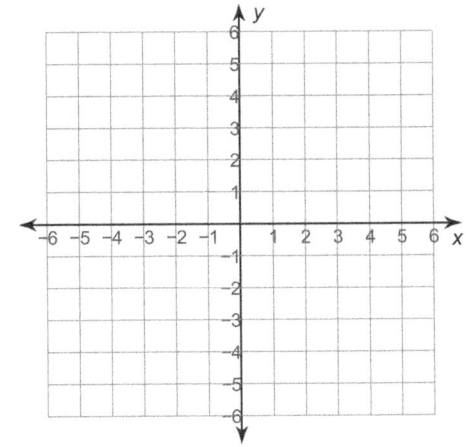
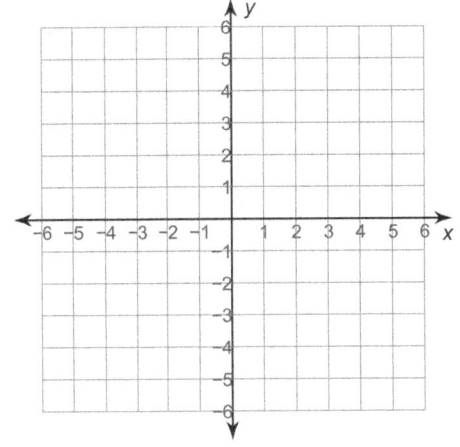

Solve the system of equations given below by using method of elimination.

(1019) $6x - 20y = -28$
   $-5x + 13y = 16$

(1020) $-55x - 10y = -10$
   $44x + 8y = 8$

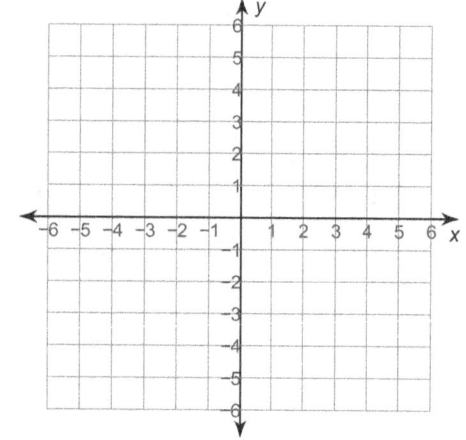
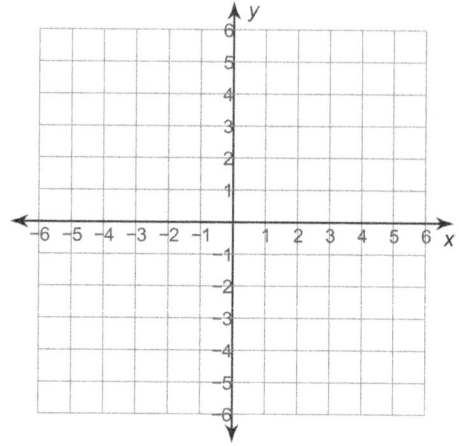

(1021) $-3x - 17y = 48$
   $19x + 16y = -29$

(1022) $2x - 5y = -57$
   $15x - 8y = 15$

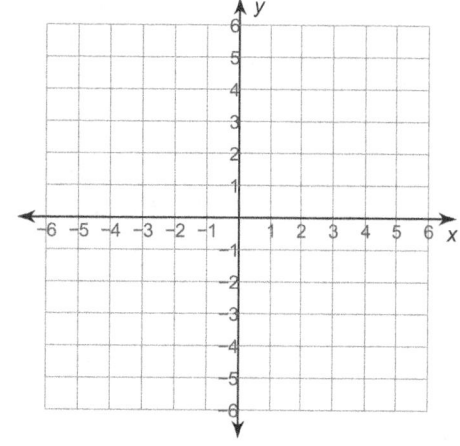
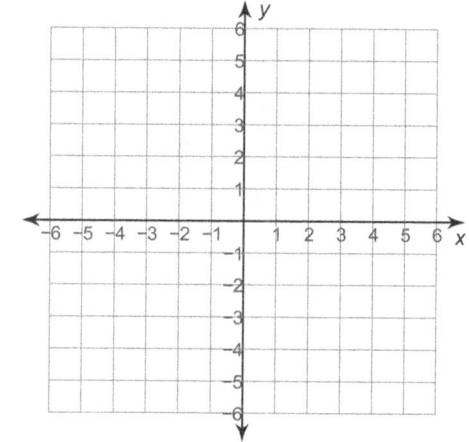

Solve the system of equations given below by using method of elimination.

(1023) $8x - 8y = 60$
$10x - 10y = 60$

(1024) $6x + 10y = -2$
$-11x - 13y = -39$

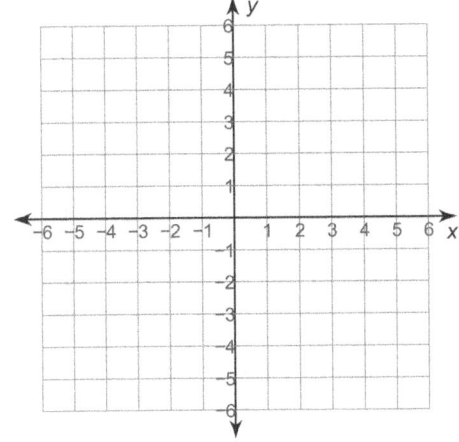
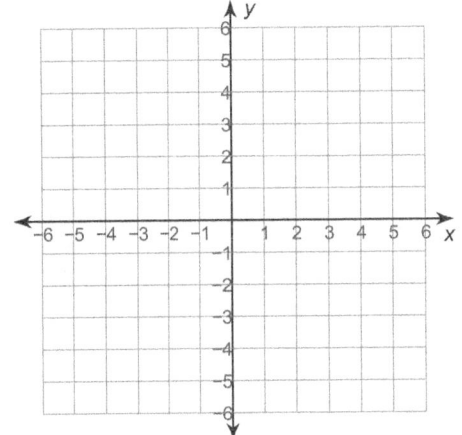

(1025) $8x - 18y = -52$
$-11x - 11y = 0$

(1026) $-18x - 12y = -30$
$-12x - 14y = 10$

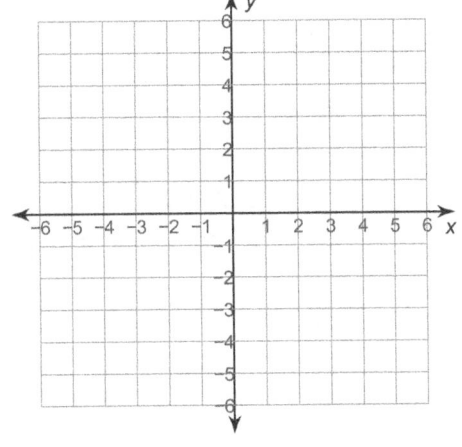
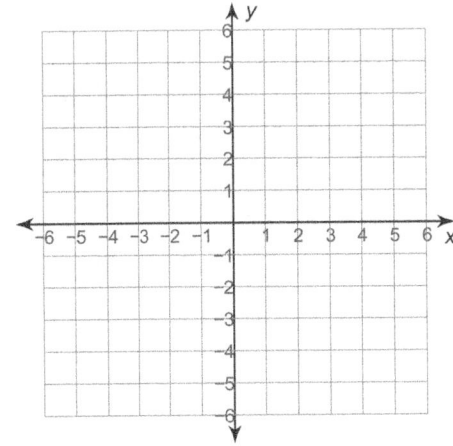

ADVANCED ALGEBRA 1

Volume 1

Solve the system of equations given below by using method of elimination.

(1027) $2x - 4y = -6$
$-13x - 3y = 39$

(1028) $13x + 19y = 42$
$16x + 17y = 7$

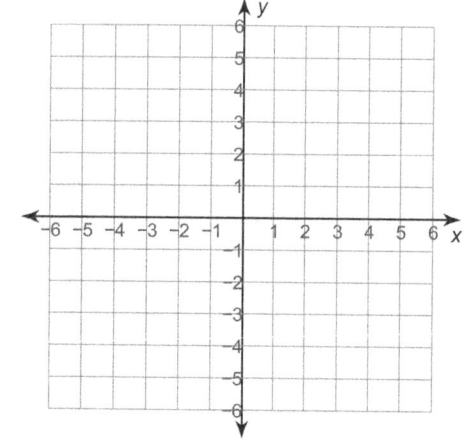
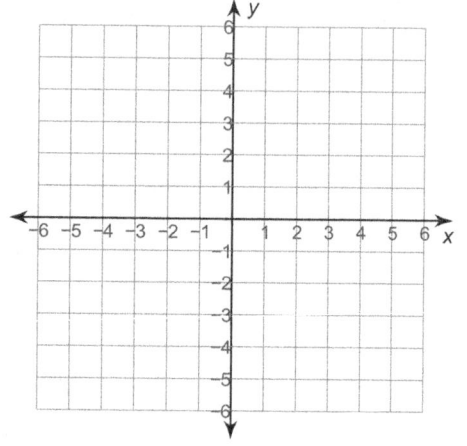

(1029) $16x - 19y = 51$
$17x + 4y = 30$

(1030) $19x + 11y = 24$
$-3x - 3y = -24$

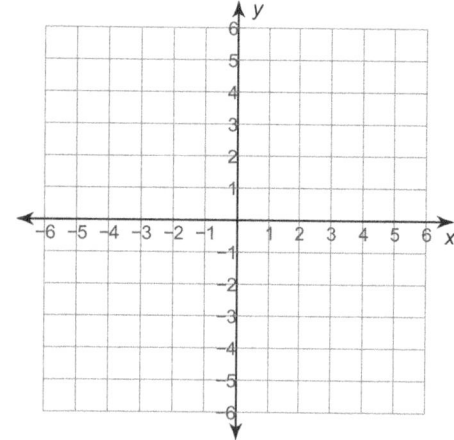
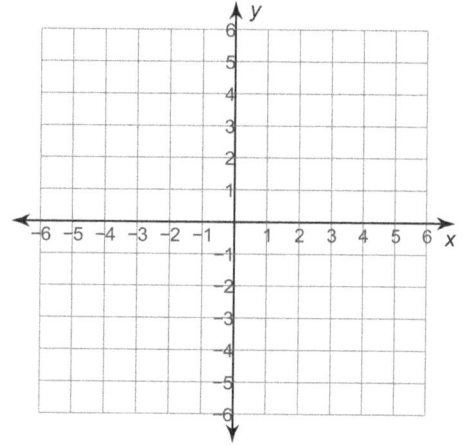

ADVANCED ALGEBRA 1

Volume 1

Solve the given system of inequalities by graphing them.

(1031) $2x + 3y > -3$
           $x - 3y > -6$

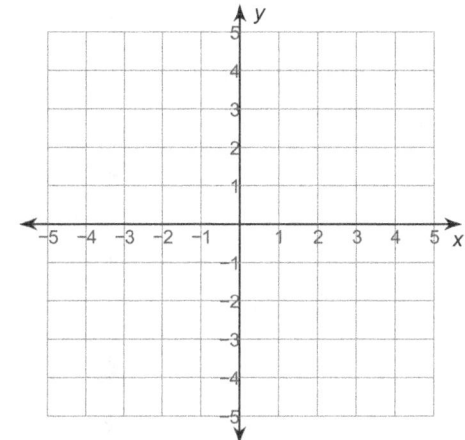

(1032) $x - y \geq 2$
           $4x + y < 3$

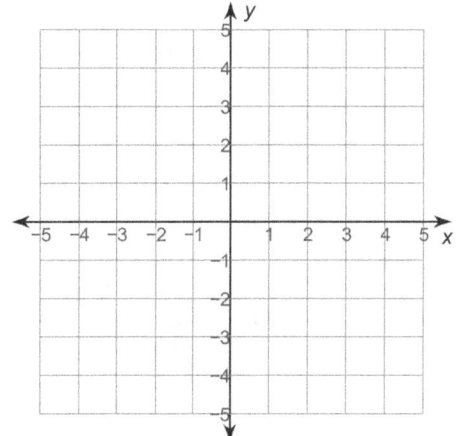

(1033) $x + 2y \leq -2$
           $3x + 2y \geq 2$

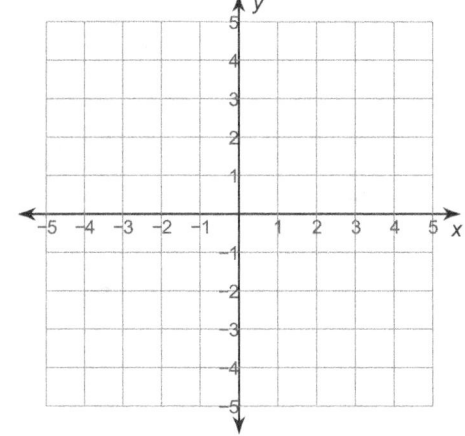

(1034) $5x + y < -2$
           $x + y > 2$

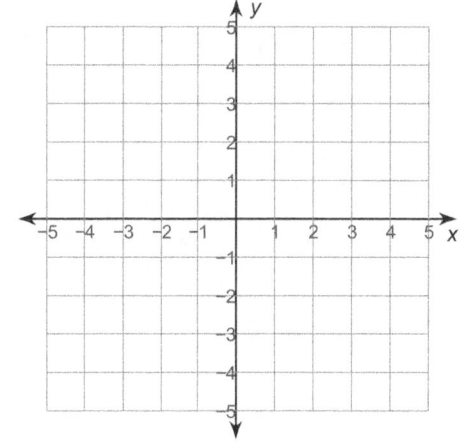

Solve the given system of inequalities by graphing them.

(1035) $x + 3y \geq 9$
$x - 3y > -3$

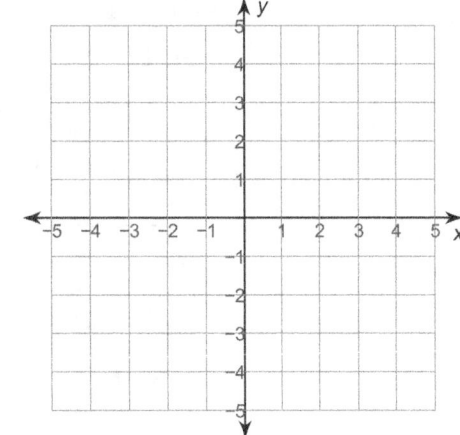

(1036) $4x + 3y \geq -3$
$x + 3y < 6$

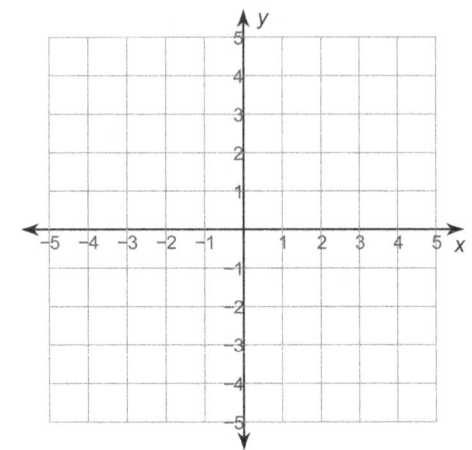

(1037) $x + 2y \geq 6$
$x - 2y \geq -2$

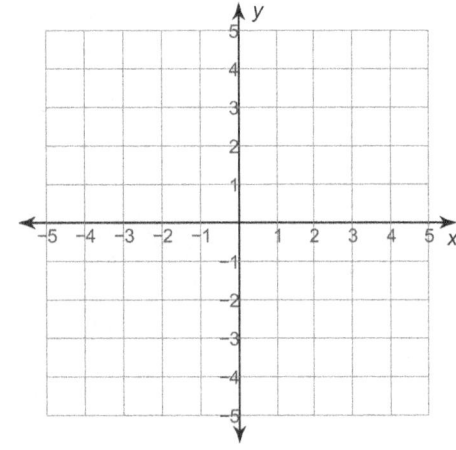

(1038) $x - y \geq 3$
$x + y > 1$

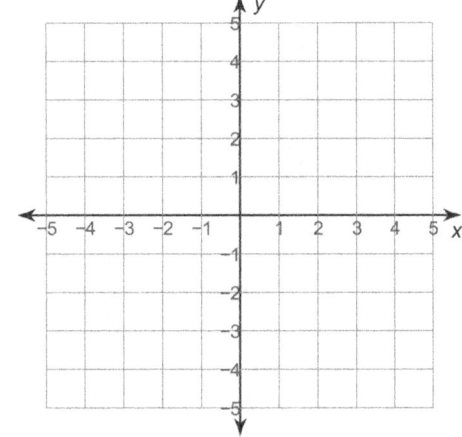

Solve the given system of inequalities by graphing them.

(1039) $x - 3y > 3$
          $x + 3y \leq -9$

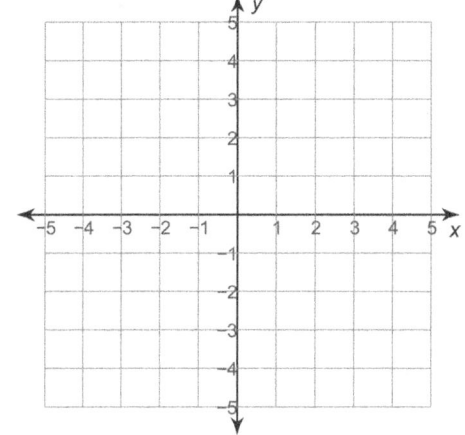

(1040) $x - 3y > -3$
          $5x - 3y \leq 9$

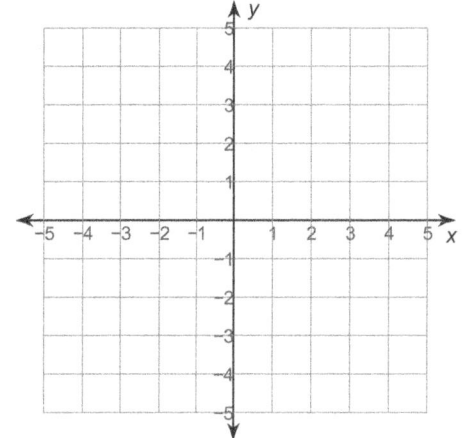

(1041) $x + 2y < 4$
          $x \geq 2$

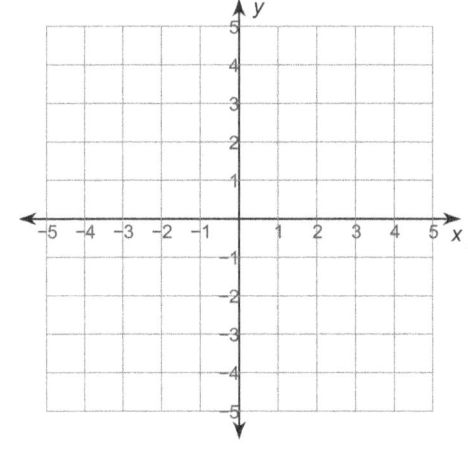

(1042) $x + 2y \geq -4$
          $2x + y < 1$

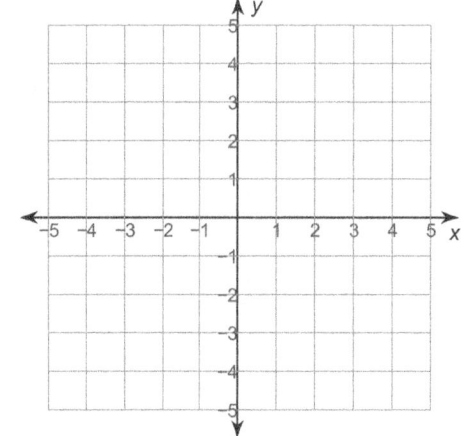

Solve the given system of inequalities by graphing them.

(1043) $2x + y \geq 3$
$2x + 3y \leq -3$

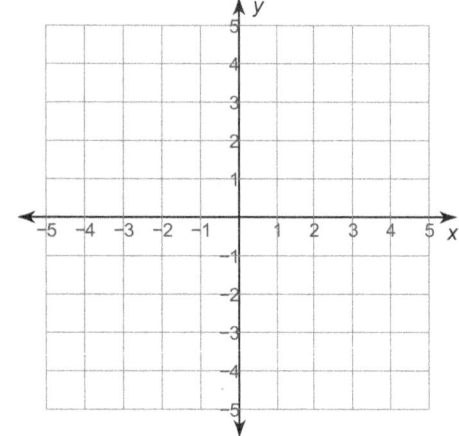

(1044) $x \leq -2$
$x - 2y \geq 2$

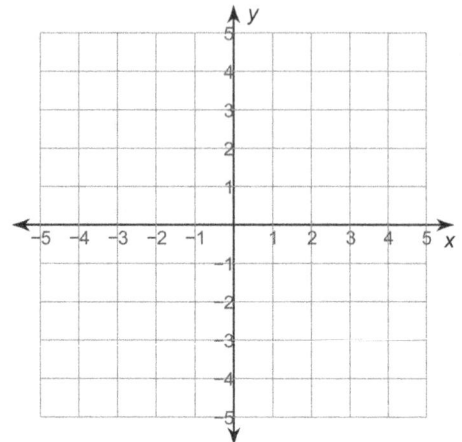

(1045) $x + y > -3$
$x - y \geq 1$

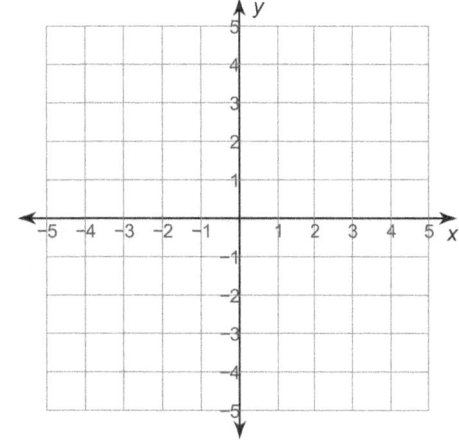

(1046) $x - 3y \leq -6$
$2x + 3y \leq -3$

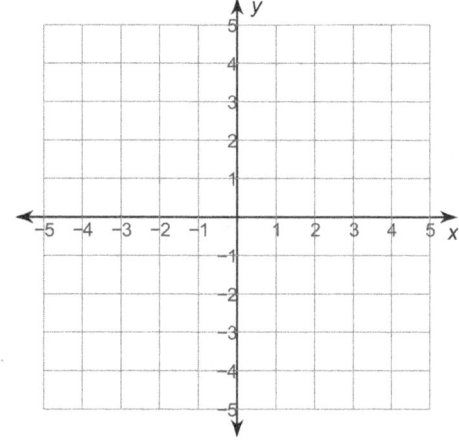

Solve the given system of inequalities by graphing them.

(1047) $x + 2y \geq 4$
        $2x + y \leq -1$

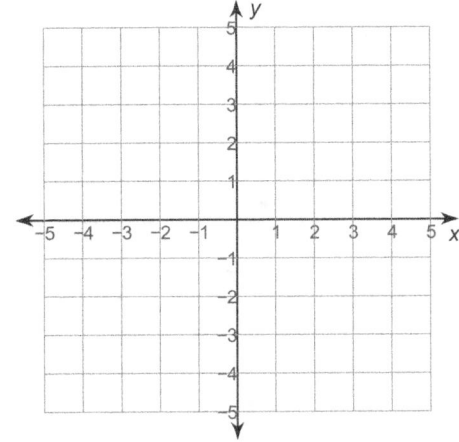

(1048) $x - y < 2$
        $5x - y \geq -2$

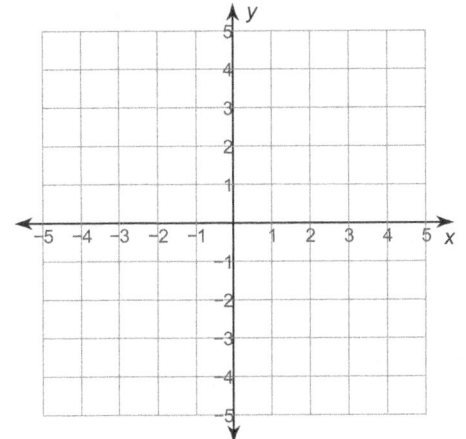

(1049) $x + y > -1$
        $x - 3y > -9$

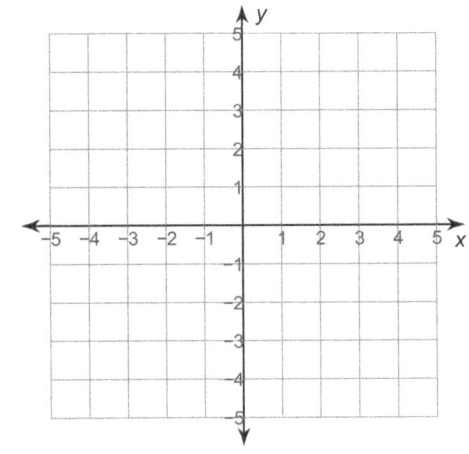

(1050) $x - y > -2$
        $x - y \leq -1$

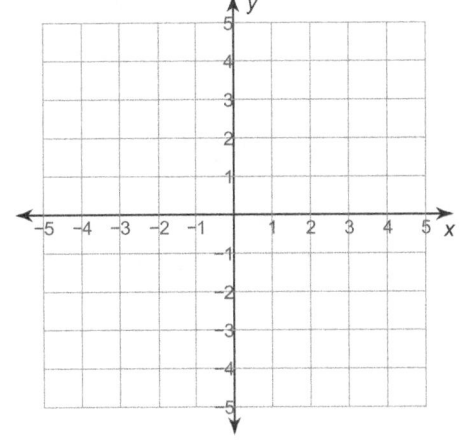

Solve the given system of inequalities by graphing them.

(1051) $x + 3y > -6$
$2x + y < 3$

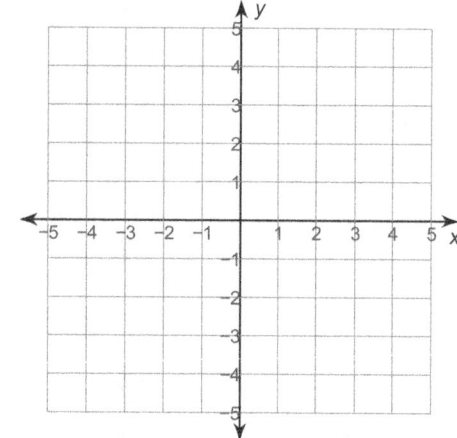

(1052) $y < -3$
$2x + y < 3$

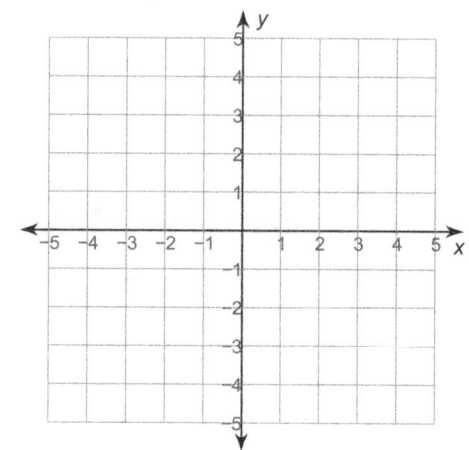

(1053) $x + 3y < -9$
$x - 3y \leq 3$

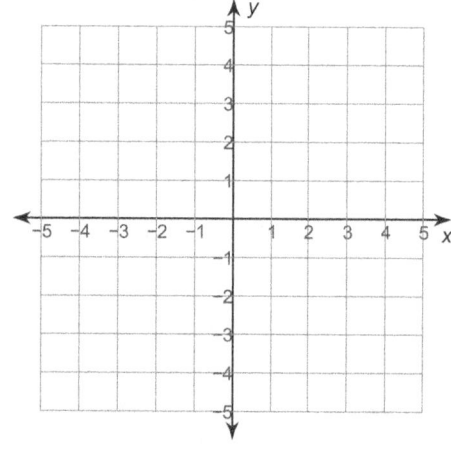

(1054) $2x + 3y \leq 9$
$4x - 3y \geq 9$

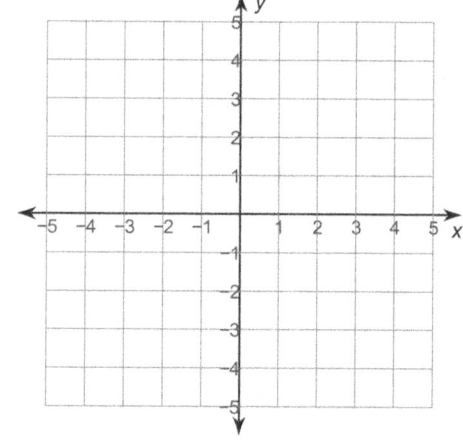

Solve the given system of inequalities by graphing them.

(1055) $x + 2y > -2$
  $x - 2y > 6$

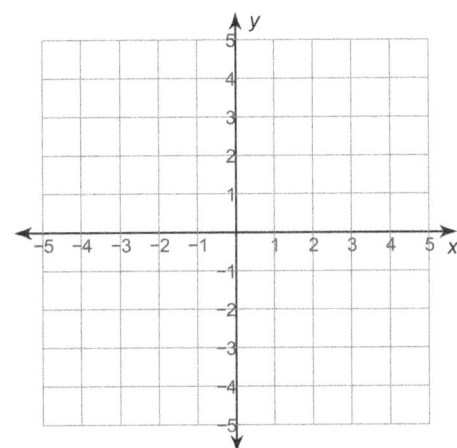

ADVANCED ALGEBRA 1

Volume 1 Answer Keys

1) $a = \dfrac{u}{k+b}$

2) $x = \dfrac{y}{uk-1}$

3) $a = \dfrac{g}{c+b}$

4) $a = \dfrac{g}{-c+b}$

5) $a = \dfrac{b}{gc-1}$

6) $x = \dfrac{u}{k+y}$

7) $a = \dfrac{b}{zm-1}$

8) $a = \dfrac{u}{k+b}$

9) $a = \dfrac{b}{zm-1}$

10) $x = \dfrac{g}{-c+y}$

11) $x = \dfrac{y}{uk-1}$

12) $x = \dfrac{y}{gc-1}$

13) $a = \dfrac{b}{gc-1}$

14) $x = \dfrac{g}{c+y}$

15) $x = \dfrac{g}{c+y}$

16) $a = \dfrac{b}{gc-1}$

17) $x = \dfrac{y}{uk-1}$

18) $a = \dfrac{u}{-k+b}$

ADVANCED ALGEBRA 1

Volume 1 Answer Keys

19) $x = \dfrac{u}{-k + y}$

20) $x = \dfrac{y}{gc - 1}$

21) $x = \dfrac{y}{uk - 1}$

22) $a = \dfrac{b}{gc - 1}$

23) $x = \dfrac{y}{zm - 1}$

24) $x = \dfrac{y}{zm - 1}$

25) $a = \dfrac{z}{m + b}$

26) $x = \dfrac{y}{uk - 1}$

27) $a = \dfrac{b}{zm - 1}$

28) $a = \dfrac{z}{-m + b}$

29) $x = \dfrac{z}{-m + y}$

30) $x = \dfrac{g}{-c + y}$

31) $x = \dfrac{p}{zn - m}$

32) $x = \dfrac{c + dr}{-y + 1}$

33) $x = \dfrac{dr}{c - y}$

34) $a = \dfrac{k + vw}{b + 1}$

35) $x = \dfrac{m - n - p}{y + 1}$

36) $x = \dfrac{m + n - p}{-y + 1}$

ADVANCED ALGEBRA 1

Volume 1 Answer Keys

37) $a = \dfrac{dr}{c-b}$

38) $x = \dfrac{z-p}{m+n}$

39) $a = \dfrac{m+np}{b-1}$

40) $a = \dfrac{m-n-p}{b-1}$

41) $x = \dfrac{m-pn}{-y-1}$

42) $x = \dfrac{k+w+v}{-y+1}$

43) $a = \dfrac{p+n}{m-b}$

44) $a = \dfrac{c-d-r}{b+1}$

45) $x = \dfrac{c-d-r}{y-1}$

46) $a = \dfrac{r}{gd-c}$

47) $a = \dfrac{k-vw}{-b-1}$

48) $a = \dfrac{m+p+n}{-b+1}$

49) $x = \dfrac{w-v}{k-y}$

50) $x = \dfrac{vw}{k-y}$

51) $a = \dfrac{z-p}{m+n}$

52) $a = \dfrac{c-d-r}{b-1}$

53) $a = \dfrac{c-d+r}{b-1}$

54) $x = \dfrac{r}{gd-c}$

55) $a = \dfrac{g-r}{c+d}$

56) $x = \dfrac{r-d}{c-y}$

57) $a = \dfrac{np}{m-b}$

58) $x = \dfrac{u}{k+w+v}$

59) $x = \dfrac{z}{m+n+p}$

60) $a = \dfrac{m+p-n}{-b+1}$

61) $a = uk$

62) $a = \dfrac{k}{u}$

63) $x = \dfrac{z}{m}$

64) $x = -u + k$

65) $x = z - m$

66) $a = gc$

67) $a = g + c$

68) $a = z + m$

69) $x = \dfrac{m}{z}$

70) $a = zm$

71) $a = \dfrac{c}{g}$

72) $x = -g + c$

ADVANCED ALGEBRA 1

73) $a = u + k$

74) $x = zm$

75) $x = \dfrac{g}{c}$

76) $x = \dfrac{u}{k}$

77) $a = -z + m$

78) $x = uk$

79) $a = \dfrac{z}{m}$

80) $x = z + m$

81) $x = -z + m$

82) $x = g + c$

83) $x = u - k$

84) $a = g - c$

85) $a = u - k$

86) $x = u + k$

87) $a = -u + k$

88) $a = -g + c$

89) $a = \dfrac{u}{k}$

90) $x = \dfrac{c}{g}$

ADVANCED ALGEBRA 1

Volume 1 Answer Keys

91) No solution.

92) {9, −9}

93) No solution.

94) No solution.

95) No solution.

96) No solution.

97) No solution.

98) {6 −6}

99) {7 −7}

100) No solution.

101) No solution.

102) No solution.

103) {8, −8}

104) {1, −1}

105) {14, −14}

106) {3, −3}

107) {4, −4}

108) {2, −2}

109) No solution.

110) No solution.

111) $\{11, -11\}$

112) $\{13, -13\}$

113) No solution.

114) No solution.

115) $\{5, -5\}$

116) No solution.

117) $\{12, -12\}$

118) No solution.

119) $\{10, -10\}$

120) $\{0\}$

121) $\{3, -3\}$

122) $\{8, -10\}$

123) $\{24, -24\}$

124) $\{36, -36\}$

125) $\{-8, 8\}$

126) $\{6, -26\}$

ADVANCED ALGEBRA 1

Volume 1 Answer Keys

127) {4, −2}   128) {0, −10}

129) {15, −9}   130) {−9, 9}

131) {−7, −9}   132) {8, −6}

133) {27, −27}   134) {9, −7}

135) {9, 5}   136) {21, −5}

137) {45, −45}   138) {10, 8}

139) {10, −10}   140) {9, 3}

141) {8, −8}   142) {−4, −8}

143) {−6, −10}   144) {4, −16}

ADVANCED ALGEBRA 1

Volume 1 Answer Keys

145) $\{-4, 4\}$ 146) $\{-8, 8\}$

147) $\{6, -10\}$ 148) $\{12, -12\}$

149) $\{2, -2\}$ 150) $\{9, -7\}$

151) $\left\{\dfrac{21}{2}, -6\right\}$ 152) $\left\{\dfrac{98}{11}, -10\right\}$

153) $\{18, -4\}$ 154) $\{-7, -17\}$

155) $\{-1, 3\}$ 156) $\left\{\dfrac{17}{3}, -11\right\}$

157) $\{1, -12\}$ 158) $\left\{\dfrac{5}{2}, -5\right\}$

159) $\left\{-\dfrac{118}{11}, 10\right\}$ 160) $\left\{-10, \dfrac{104}{9}\right\}$

161) $\{-1, -11\}$ 162) $\left\{10, -\dfrac{25}{2}\right\}$

ADVANCED ALGEBRA 1

Volume 1 Answer Keys

163) $\{12, -9\}$ 164) $\{-2, -4\}$

165) $\left\{2, -\dfrac{20}{9}\right\}$ 166) $\{2, -12\}$

167) $\{11, -7\}$ 168) $\left\{-5, \dfrac{27}{4}\right\}$

169) $\{11, -9\}$ 170) $\{30, -10\}$

171) $\left\{-\dfrac{41}{5}, 11\right\}$ 172) $\left\{9, -\dfrac{26}{3}\right\}$

173) $\left\{-2, \dfrac{21}{5}\right\}$ 174) $\left\{3, -\dfrac{23}{4}\right\}$

175) $\left\{\dfrac{91}{11}, -9\right\}$ 176) $\{3, -2\}$

177) $\left\{-\dfrac{3}{4}, 3\right\}$ 178) $\left\{\dfrac{46}{9}, -4\right\}$

179) $\left\{-11, \dfrac{51}{5}\right\}$ 180) $\left\{\dfrac{8}{3}, -4\right\}$

181) {5, −43}

182) {0}

183) {18, −18}

184) {20, −20}

185) {−3, −27}

186) {13}

187) {56, −18}

188) No solution.

189) No solution.

190) {17, −17}

191) {20, −20}

192) {20, −20}

193) {10, −10}

194) {18, −18}

195) {8, −28}

196) $\left\{\dfrac{119}{20}, -\dfrac{119}{20}\right\}$

197) No solution.

198) {15, −15}

199) {57, −97}

200) {14, −14}

201) No solution.

202) No solution.

203) No solution.

204) {23, −5}

205) {−2, 2}

206) {−8, −32}

207) {3, −3}

208) $\left\{-\dfrac{5}{8}, \dfrac{5}{8}\right\}$

209) {48, −8}

210) {2, −2}

211) $\left\{6, -\dfrac{76}{13}\right\}$

212) {−16, 18}

213) {−5, 13}

214) $\left\{17, -\dfrac{153}{7}\right\}$

215) $\left\{\dfrac{2}{11}, 0\right\}$

216) $\left\{0, \dfrac{34}{13}\right\}$

ADVANCED ALGEBRA 1

Volume 1 Answer Keys

217) $\{5, -7\}$

218) $\left\{0, -\dfrac{5}{2}\right\}$

219) $\left\{0, -\dfrac{6}{7}\right\}$

220) $\left\{2, -\dfrac{30}{7}\right\}$

221) $\left\{15, -\dfrac{37}{2}\right\}$

222) $\left\{14, -\dfrac{59}{3}\right\}$

223) $\left\{0, -\dfrac{28}{11}\right\}$

224) $\left\{-1, -\dfrac{9}{5}\right\}$

225) $\{2, -4\}$

226) $\left\{5, -\dfrac{79}{15}\right\}$

227) $\{-7, 19\}$

228) $\{-7, 13\}$

229) $\left\{-3, \dfrac{53}{15}\right\}$

230) $\left\{0, -\dfrac{4}{5}\right\}$

231) $\left\{-\dfrac{22}{7}, 6\right\}$

232) $\left\{-\dfrac{29}{13}, 5\right\}$

233) $\left\{\dfrac{5}{3}, -4\right\}$

234) $\left\{\dfrac{34}{3}, -14\right\}$

235) $\{2, -7\}$

236) $\left\{0, \dfrac{38}{15}\right\}$

237) $\left\{\dfrac{25}{13}, -1\right\}$

238) $\{17, -7\}$

239) $\left\{2, -\dfrac{6}{19}\right\}$

240) $\left\{0, \dfrac{2}{5}\right\}$

241) $\left\{5, -\dfrac{3}{7}\right\}$

242) $\left\{\dfrac{54}{5}, -11\right\}$

243) $\left\{\dfrac{29}{17}, -1\right\}$

244) $\left\{2, \dfrac{12}{13}\right\}$

245) $\left\{-\dfrac{59}{15}, 3\right\}$

246) $\left\{-13, \dfrac{35}{4}\right\}$

247) $\{5, -1\}$

248) $\left\{-\dfrac{1}{11}, 3\right\}$

249) $\left\{\dfrac{8}{15}, 0\right\}$

250) $\left\{-\dfrac{13}{17}, -1\right\}$

251) $\left\{\dfrac{4}{3}, -3\right\}$

252) $\left\{\dfrac{82}{9}, -9\right\}$

253) $\left\{3, -\dfrac{5}{7}\right\}$

254) $\left\{\dfrac{29}{6}, -5\right\}$

255) $\left\{\dfrac{13}{8}, -1\right\}$

256) $\left\{-1, -\dfrac{3}{4}\right\}$

257) $\left\{7, -\dfrac{1}{3}\right\}$

258) $\left\{1, \dfrac{1}{7}\right\}$

259) $\left\{-2, -\dfrac{12}{13}\right\}$

260) $\left\{-\dfrac{1}{17}, 1\right\}$

261) $n \geq 3$ or $n \leq -3$: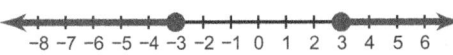

262) $-7 \leq k \leq 7$: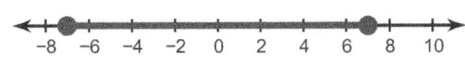

263) $-10 < r < 18$: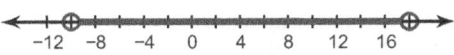

264) $a \geq 10$ or $a \leq -10$: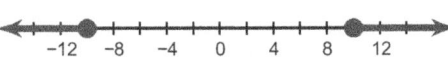

265) $-9 < v < -7$: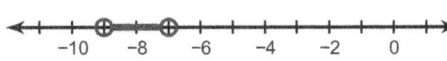

266) $b > 19$ or $b < 1$:

267) $x \geq 10$ or $x \leq -10$:

268) $n > 6$ or $n < -6$:

269) $n < -3$ or $n > 3$:

270) $-7 < a < 15$:

271) $-10 \leq k \leq 10$:

272) $n > 5$ or $n < -21$:

273) $-3 < m < 21$:

274) $a \geq -2$ or $a \leq -6$:

275) $a > 3$ or $a < -3$:

276) $2 < x < 6$:

277) $1 \leq v \leq 5$:

278) $-16 < r < 6$:

279) $-18 < m < 0$:

280) $-8 < b < 8$:

281) $-11 < b < 3$:

282) $-8 \leq b \leq 8$:

283) $-10 \leq b \leq 10$:

284) $a \geq 23$ or $a \leq -9$:

285) $a > 3$ or $a < -21$:

286) $5 < n < 13$:

287) $k > 31$ or $k < -33$:

288) $-9 \leq a \leq 29$:

289) $r < -4$ or $r > 4$:

290) $p \geq 8$ or $p \leq -8$:

291) $0 < x < 3$:

292) $1 \leq x \leq 13$:

293) $r > 0$ or $r < -\dfrac{7}{5}$:

294) $x \geq \dfrac{7}{4}$ or $x \leq -1$:

295) $x < -6$ or $x > 2$:

296) $0 \leq x \leq \dfrac{14}{9}$:

297) $x < -\dfrac{7}{3}$ or $x > 7$:

298) $v > 0$ or $v < -\dfrac{18}{7}$:

299) $-3 < n < 1$:

300) $-\dfrac{23}{4} < n < 4$:

301) $-3 < k < \dfrac{3}{2}$:

ADVANCED ALGEBRA 1

Volume 1 Answer Keys

302) $x > 0$ or $x < -\dfrac{3}{4}$:

303) $m \geq 5$ or $m \leq -23$:

304) $a > 1$ or $a < -\dfrac{13}{5}$:

305) $m \geq 0$ or $m \leq -\dfrac{4}{3}$:

306) $x < -\dfrac{25}{9}$ or $x > 1$:

307) $n \geq -3$ or $n \leq -\dfrac{11}{3}$:

308) $x \geq 3$ or $x \leq 0$:

309) $n \geq \dfrac{26}{5}$ or $n \leq -2$:

310) $-8 < p < \dfrac{4}{3}$:

ADVANCED ALGEBRA 1

Volume 1 Answer Keys

311) $-2 < r < \dfrac{6}{5}$:

312) $x < 0$ or $x > 6$:

313) $0 \le x \le \dfrac{8}{5}$:

314) $-\dfrac{7}{4} \le k \le 0$:

315) $n > -\dfrac{6}{7}$ or $n < -2$:

316) $x \le -6$ or $x \ge 7$:

317) $k < 0$ or $k > 2$:

318) $-\dfrac{8}{5} \le r \le 4$:

319) $n < -1$ or $n > \dfrac{17}{9}$:

 **ADVANCED ALGEBRA 1**

320) $-2 \le x \le 1$:

321) $\dfrac{2}{5}$

322) Undefined

323) $-\dfrac{1}{6}$

324) Undefined

325) $\dfrac{1}{6}$

326) 4

327) 1

328) 2

329) 4

330) -2

331) $\dfrac{2}{3}$

332) $-\dfrac{3}{5}$

333) 1

334) $\dfrac{2}{3}$

335) $-\dfrac{3}{5}$

336) $\dfrac{6}{5}$

ADVANCED ALGEBRA 1

337) $-\dfrac{1}{2}$ 338) $-\dfrac{5}{3}$

339) $\dfrac{1}{2}$ 340) 3

341) $\dfrac{1}{5}$ 342) 1

343) -1 344) 1

345) -4 346) $\dfrac{2}{7}$

347) $\dfrac{6}{11}$ 348) $-\dfrac{7}{6}$

349) $\dfrac{21}{2}$ 350) $-\dfrac{1}{7}$

351) $\dfrac{28}{25}$ 352) $\dfrac{1}{4}$

353) $\dfrac{1}{5}$ 354) $\dfrac{1}{5}$

ADVANCED ALGEBRA 1

Volume 1 Answer Keys

355) $\dfrac{21}{4}$ 356) $\dfrac{28}{3}$

357) $\dfrac{16}{21}$ 358) $-\dfrac{4}{23}$

359) $-\dfrac{7}{9}$ 360) $-\dfrac{7}{10}$

361) $-\dfrac{10}{9}$ 362) $\dfrac{20}{11}$

363) $\dfrac{17}{5}$ 364) $\dfrac{3}{5}$

365) -1 366) $\dfrac{2}{3}$

367) $-\dfrac{5}{3}$ 368) $-\dfrac{6}{29}$

369) $\dfrac{3}{4}$ 370) $\dfrac{4}{5}$

371) 3 372) 2 373) $\dfrac{2}{3}$

Volume 1 Answer Keys

374) −2

375) 0

376) 1

377) $-\dfrac{5}{3}$

378) Undefined

379) 0

380) Undefined

381) 7

382) $-\dfrac{5}{2}$

383) $-\dfrac{4}{3}$

384) Undefined

385) 1

386) $\dfrac{3}{2}$

387) 0

388) $-\dfrac{2}{5}$

389) 10

390) 0

391) 6

392) $-\dfrac{7}{3}$

393) -1

394) $\dfrac{1}{2}$

395) Undefined

396) $-\dfrac{2}{3}$

397) 5

398) 1

399) $\dfrac{5}{4}$

400) $-\dfrac{5}{3}$

401) -1

402) $\dfrac{3}{5}$

403) 0

404) $-\dfrac{8}{5}$

405) $\dfrac{4}{3}$

406) 3

407) $-\dfrac{1}{2}$

408) $\dfrac{2}{3}$

409) $-\dfrac{2}{3}$

410) 2

411) 1

412) −2

413) $\dfrac{8}{3}$

414) 2

415) 1

416) $-\dfrac{1}{4}$

417) −5

418) $-\dfrac{5}{3}$

419) 7

420) 2

421) −2

422) $\dfrac{1}{4}$

423) −4

424) $\dfrac{2}{3}$

425) $\dfrac{9}{2}$

426) $\dfrac{3}{2}$

427) $-\dfrac{1}{2}$

ADVANCED ALGEBRA 1

Volume 1 Answer Keys

428) −1

429) 3

430) −5

431) $\dfrac{1}{2}$

432) $-\dfrac{5}{4}$

433) $\dfrac{2}{5}$

434) 0

435) $-\dfrac{1}{5}$

436) $-\dfrac{1}{6}$

437) Undefined

438) 2

439) Undefined

440) $-\dfrac{5}{2}$

441) $-\dfrac{2}{3}$

442) $\dfrac{1}{5}$

443) −4

444) 5

445) $-\dfrac{5}{8}$

446) $-\dfrac{5}{6}$

447) $-\dfrac{5}{6}$

448) $\dfrac{5}{8}$

449) $\dfrac{1}{2}$

450) $\dfrac{1}{3}$

451) $\dfrac{4}{3}$

452) 0

453) 0

454) 1

455) $-\dfrac{3}{2}$

456) 35

457) 32

458) 0

459) −30

460) −46

461) 11

462) 1

463) −23

ADVANCED ALGEBRA 1

Volume 1 Answer Keys

464) −34   465) 23

466) −37   467) −24

468) −5   469) 21

470) 38   471) 5

472) 18   473) 41

474) −33   475) 18

476) −44   477) 41

478) 26   479) −29

480) −49   481) −7

482) 5

483) 34

484) 35

485) 8

486) $3x + 5y = 0$

487) $2x - 3y = 6$

488) $x + 2y = 4$

489) $x - 3y = 9$

490) $7x + 3y = 6$

491) $2x + 3y = 15$

492) $5x - 3y = -12$

493) $3x + 2y = 6$

494) $9x + y = -4$

495) $5x - y = 0$

496) $2x - y = -3$

497) $3x + y = -3$

498) $2x - 5y = 10$

499) $9x - 5y = -25$

500) $y = -5$

501) $4x + y = 2$

502) $x + y = 2$

503) $x - 2y = 4$

504) $5x - 3y = 6$

505) $5x + 2y = -4$

506) $2x - y = 1$

507) $3x + 5y = -15$

508) $6x + 5y = -10$

509) $3x - 5y = -20$

510) $8x + 5y = -20$

511) $y = \dfrac{1}{2}x - 2$

512) $y = -3x$

513) $y = \dfrac{1}{5}x$

514) $y = -\dfrac{1}{5}x - 4$

515) $y = -2$

516) $y = \dfrac{3}{5}x - 1$

517) $y = -\dfrac{2}{5}x + 3$

ADVANCED ALGEBRA 1

Volume 1
Answer Keys

518) $y = -3x + 4$

519) $y = -8x + 4$

520) $y = \dfrac{1}{5}x - 1$

521) $y = 4x - 4$

522) $y = -\dfrac{1}{5}x - 5$

523) $y = x - 2$

524) $y = \dfrac{1}{5}x + 2$

525) $y = \dfrac{8}{3}x - 5$

526) $y = x + 4$

527) $y = 2x + 3$

528) $y = 7x + 5$

529) $y = -\dfrac{7}{4}x + 5$

530) $y = -6x + 1$

531) $y = \dfrac{5}{3}x - 1$

532) $y = \dfrac{1}{5}x - 3$

533) $y = \dfrac{7}{4}x - 4$

534) $y = -\dfrac{1}{2}x$

535) $y = 4x - 1$

536) $5x - y = -1$

537) $3x + 4y = -4$

538) $3x + y = 1$

539) $4x + 3y = -6$

540) $5x - y = 0$

541) $7x + 3y = 6$

542) $4x - 5y = -20$

543) $x = 3$

544) $7x - y = 2$

545) $6x + y = 3$

546) $3x - 4y = -8$

547) $7x + 3y = -12$

548) $x + 2y = -6$

549) $4x - 3y = 1$

550) $x + y = -1$

551) $7x - 2y = 10$

552) $x = -3$

553) $4x - y = 5$

554) $5x - y = -5$

555) $x - 9y = -13$

556) $x - y = -4$

557) $3x - y = -7$

558) $2x + y = -6$

559) $x + 3y = 15$

560) $7x - 2y = 20$

561) $y = 6x + 4$

562) $y = \dfrac{3}{5}x + 5$

563) $y = \dfrac{2}{5}x - 2$

564) $y = \dfrac{2}{3}x - 2$

565) $y = \dfrac{7}{5}x + 5$

566) $y = -x$

567) $y = \dfrac{4}{3}x - 4$

568) $y = \dfrac{5}{2}x + 10$

569) $y = 2x - 5$

570) $x = -4$

571) $x = 5$

ADVANCED ALGEBRA 1

572) $y = -\dfrac{9}{5}x - 5$

573) $y = -\dfrac{1}{2}x + 2$

574) $y = 5x + 2$

575) $y = \dfrac{2}{3}x + 2$

576) $y = \dfrac{7}{5}x - 3$

577) $y = 3x - 4$

578) $x = -2$

579) $y = -5x + 5$

580) $y = 3x + 2$

581) $y = \dfrac{6}{5}x - 5$

582) $y = -x - 4$

583) $x = -3$

584) $y = -\dfrac{1}{4}x - 5$

585) $y = -\dfrac{3}{2}x - 3$

586) $0 = x + 2$

587) $y - 5 = -9x$

588) $y + 4 = \dfrac{4}{3}(x + 4)$

589) $y + 5 = -5(x - 2)$

ADVANCED ALGEBRA 1

Volume 1 Answer Keys

590) $y + 2 = -2(x - 3)$

591) $y + 4 = 0$

592) $y - 4 = 3(x - 2)$

593) $y + 5 = 2(x + 2)$

594) $0 = x - 2$

595) $y = -\frac{5}{7}(x + 3)$

596) $y - 4 = -(x - 1)$

597) $y + 5 = \frac{5}{4}(x + 4)$

598) $0 = x - 1$

599) $y - 4 = 3(x - 1)$

600) $y - 5 = -9(x + 1)$

601) $y + 1 = -\frac{1}{2}(x + 2)$

602) $y - 5 = \frac{9}{2}(x - 2)$

603) $y + 4 = \frac{5}{4}(x + 4)$

604) $y - 4 = -2(x + 4)$

605) $y + 2 = -\frac{7}{2}x$

606) $y - 4 = -\frac{4}{3}x$

607) $y + 3 = 2(x + 3)$

ADVANCED ALGEBRA 1

Volume 1 Answer Keys

608) $y + 2 = \frac{2}{5}(x - 5)$

609) $y + 1 = -\frac{1}{4}(x + 4)$

610) $y + 2 = -\frac{1}{3}(x - 3)$

611) $y = 5x + 10$

612) $y = 5x - 5$

613) $y = -\frac{2}{3}x + \frac{16}{3}$

614) $y = 3x - 3$

615) $y = \frac{4}{5}x + 1$

616) $y = -\frac{1}{2}x - 5$

617) $y = \frac{1}{2}x - 4$

618) $y = \frac{6}{5}x + \frac{4}{5}$

619) $y = 2x - 3$

620) $y = -x + 5$

621) $y = x - 2$

622) $y = \frac{3}{4}x - 3$

623) $y = -\frac{7}{2}x - \frac{15}{2}$

624) $y = -4x - 10$

625) $y = \frac{2}{5}x - \frac{7}{5}$

626) $y = -\dfrac{2}{5}x + \dfrac{16}{5}$

627) $y = -\dfrac{2}{3}x - \dfrac{1}{3}$

628) $y = \dfrac{3}{7}x + \dfrac{20}{7}$

629) $y = -\dfrac{1}{3}x + 2$

630) $y = -\dfrac{1}{3}x - \dfrac{5}{3}$

631) $y = -\dfrac{7}{6}x + \dfrac{5}{3}$

632) $y = \dfrac{1}{4}x + 4$

633) $y = -2x + 1$

634) $y = -x + 4$

635) $y = -\dfrac{7}{4}x + 4$

636) $y = \dfrac{3}{5}x - \dfrac{2}{5}$

637) $y = x$

638) $y = \dfrac{7}{3}x - 5$

639) $y = -\dfrac{7}{5}x - 5$

640) $y = \dfrac{6}{5}x - 5$

641) $x = -2$

642) $y = -4$

643) $y = x - 1$

644) $y = -\dfrac{1}{2}x - 5$

645) $y = 3x - 4$

646) $y = -\dfrac{4}{7}x - \dfrac{27}{7}$

647) $y = -\dfrac{4}{3}x - 4$

648) $y = 2x - 4$

649) $y = -\dfrac{6}{5}x - 5$

650) $y = \dfrac{5}{4}x - 1$

651) $y = -\dfrac{3}{2}x$

652) $x = 2$

653) $y = -\dfrac{1}{4}x + 2$

654) $y = -\dfrac{3}{2}x + 1$

655) $y = -2x + 5$

656) $y = 2x - 5$

657) $y = -\dfrac{7}{5}x + 5$

658) $y = \dfrac{1}{3}x + 1$

659) $y = x + 2$

660) $y = -x + 3$

661) $y = 4x - 3$

ADVANCED ALGEBRA 1

Volume 1
Answer Keys

662) $y = \dfrac{1}{3}x - 5$

663) $y = -\dfrac{3}{2}x + 2$

664) $y = -x + 7$

665) $y = \dfrac{3}{4}x - 5$

666) $y - 5 = \dfrac{7}{3}(x - 3)$

667) $y - 2 = -\dfrac{1}{3}(x + 3)$

668) $y + 4 = \dfrac{9}{5}(x + 5)$

669) $y + 1 = \dfrac{3}{4}(x + 3)$

670) $y - 5 = 4(x + 2)$

671) $y - 5 = -2x$

672) $y - 1 = -\dfrac{1}{4}(x - 4)$

673) $y + 4 = \dfrac{1}{5}(x + 5)$

674) $y - 1 = -\dfrac{3}{5}(x - 5)$

675) $y + 3 = -\dfrac{3}{4}(x - 2)$

676) $y - 5 = -\dfrac{9}{5}(x + 5)$

677) $y = -4(x - 1)$

678) $y - 5 = 10(x - 1)$

679) $y - 3 = \dfrac{1}{2}(x - 4)$

680) $y + 5 = \dfrac{2}{3}(x + 3)$

681) $y + 4 = -3(x - 1)$

682) $y + 1 = \dfrac{1}{3}(x - 3)$

683) $y = \dfrac{2}{3}(x - 2)$

684) $y + 2 = -\dfrac{5}{6}(x - 4)$

685) $y + 5 = 4(x + 1)$

686) $y - 5 = -\dfrac{1}{3}(x + 3)$

687) $y = -\dfrac{3}{4}(x + 4)$

688) $y - 4 = 6(x - 1)$

689) $y - 4 = \dfrac{1}{5}(x - 5)$

690) $y + 4 = \dfrac{7}{4}(x + 4)$

691) $x + y = -2$

692) $x - 2y = -2$

693) $x - 3y = 5$

694) $3x - 2y = -10$

695) $x - 5y = 10$

696) $x + 10y = 5$

697) $6x - 5y = 15$

698) $y = 2$

699) $5x + 3y = 9$

700) $y = 5$

701) $2x + 5y = 0$

702) $3x + 2y = 6$

703) $2x + y = 2$

704) $4x + y = 15$

705) $x = 3$

706) $x + y = 4$

707) $8x - 7y = -12$

708) $x - y = -3$

709) $8x - 3y = -15$

710) $x - y = -1$

711) $8x - 3y = 15$

712) $3x + 5y = 10$

713) $3x + y = 5$

714) $x - y = 2$

715) $6x + 5y = -15$

716) $x - y = 0$

717) $3x - y = 4$

718) $x + y = 0$

719) $7x + 3y = -15$

720) $2x + y = 3$

721) $2x - y = 1$

722) $5x - 4y = 16$

723) $2x + 5y = 25$

724) $x - 3y = -6$

725) $7x - 2y = 6$

726) $x - 2y = -4$

727) $5x - y = -7$

728) $6x - y = 28$

729) $3x + 2y = 15$

730) $6x + y = -4$

731) $x + y = -2$

732) $3x + 2y = -4$

ADVANCED ALGEBRA 1

Volume 1 Answer Keys

733) $y = -3$ 734) $3x + 5y = -20$

735) $3x - 4y = -12$ 736) $3x + 2y = -3$

737) $x + y = 5$ 738) $7x - y = -5$

739) $x = 0$ 740) $x = -2$

741) $y = \dfrac{1}{4}x - 4$ 742) $y = -2x + 8$

743) $y = 4x - 5$ 744) $y = 9x - 4$

745) $y = \dfrac{1}{3}x - 4$ 746) $y = -\dfrac{3}{5}x$

747) $y = \dfrac{4}{3}x - 5$ 748) $y = -x$

749) $y = -\dfrac{8}{5}x - 4$ 750) $y = -2x + 3$

227

ADVANCED ALGEBRA 1

Volume 1
Answer Keys

751) $y = \dfrac{1}{5}x + 3$

752) $y = 4x - 1$

753) $y = 4x - 4$

754) $y = -\dfrac{5}{3}x - 1$

755) $y = -\dfrac{5}{2}x + \dfrac{15}{2}$

756) $y = -\dfrac{1}{5}x$

757) $y = \dfrac{9}{4}x + 4$

758) $y = \dfrac{9}{2}x - 4$

759) $y = \dfrac{9}{5}x + \dfrac{11}{5}$

760) $y = \dfrac{3}{5}x - 3$

761) $y = \dfrac{1}{5}x + 5$

762) $y = -\dfrac{5}{3}x - \dfrac{10}{3}$

763) $y = -\dfrac{1}{2}x + 3$

764) $y = \dfrac{1}{2}x - \dfrac{3}{2}$

765) $y = \dfrac{1}{4}x + 2$

766) $y - 4 = 5(x - 1)$

767) $y + 4 = -(x + 3)$

768) $y + 3 = 2(x + 3)$

ADVANCED ALGEBRA 1

Volume 1
Answer Keys

769) $y - 3 = -\dfrac{1}{2}(x - 2)$

770) $y - 5 = -3(x + 2)$

772) $y + 3 = -7(x + 4)$

771) $y + 2 = 0$

773) $y + 1 = \dfrac{1}{2}(x + 2)$

774) $y + 2 = -3(x + 2)$

775) $y - 2 = -\dfrac{2}{3}(x - 4)$

776) $y - 3 = -\dfrac{5}{2}(x + 5)$

777) $y - 1 = -\dfrac{1}{3}(x - 1)$

778) $y = -x$

779) $y - 1 = \dfrac{4}{5}(x + 5)$

780) $y + 5 = 8(x + 1)$

781) $y - 3 = -(x + 5)$

782) $y - 1 = -\dfrac{1}{2}(x + 4)$

783) $y - 3 = -\dfrac{1}{2}(x + 5)$

784) $y + 1 = 3(x + 5)$

785) $0 = x - 3$

786) $y - 3 = 2(x + 1)$

787) $y - 5 = 2(x - 2)$

788) $y - 5 = x - 4$

789) $y + 5 = 2(x + 2)$

790) $y = 0$

791)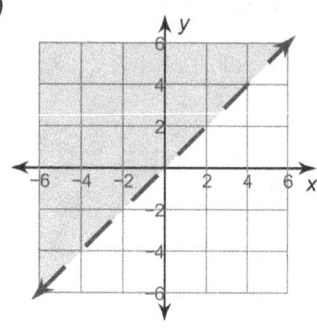

792)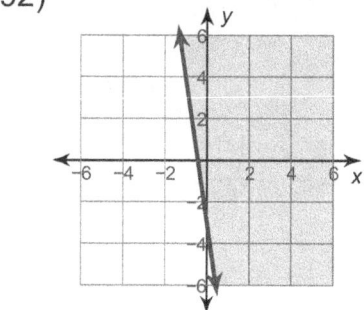

793)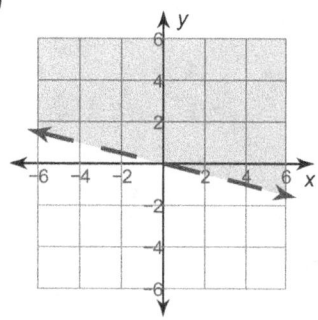

794)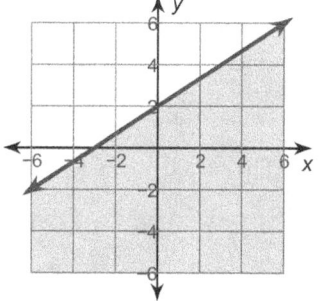

795)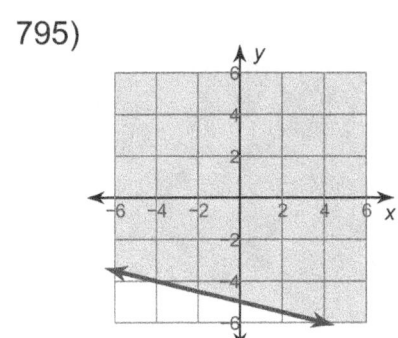

796)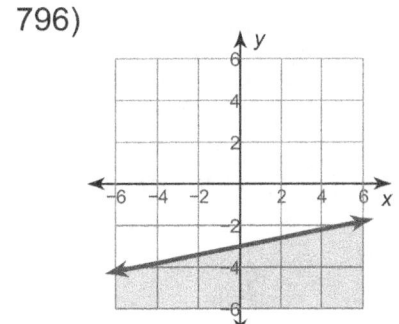

797)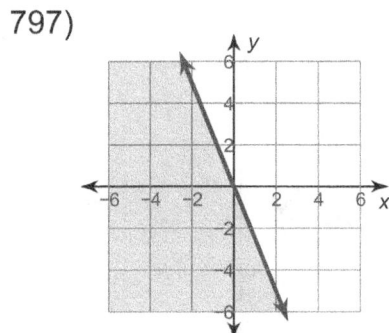

798)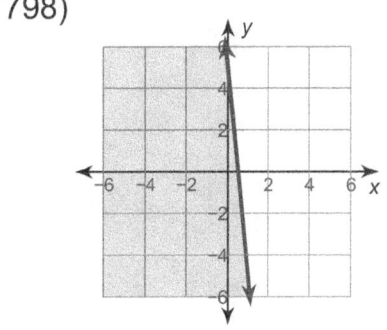

799)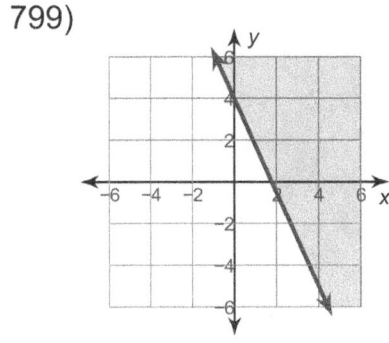

800)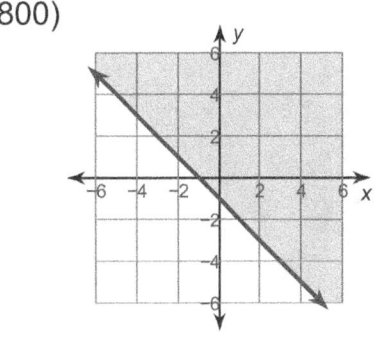

801)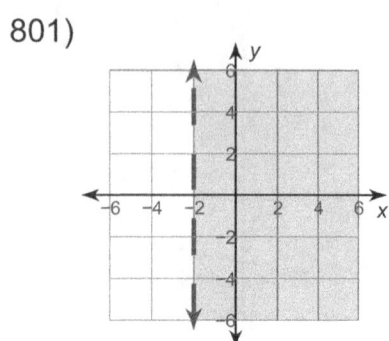

802)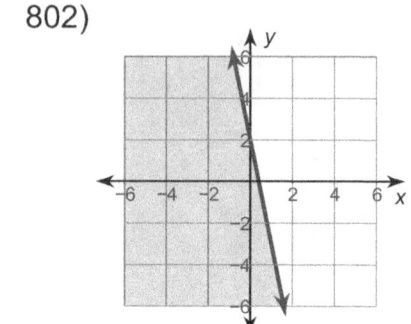

803)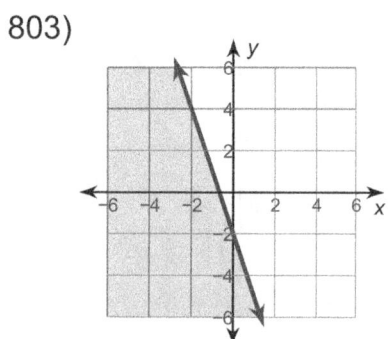

804)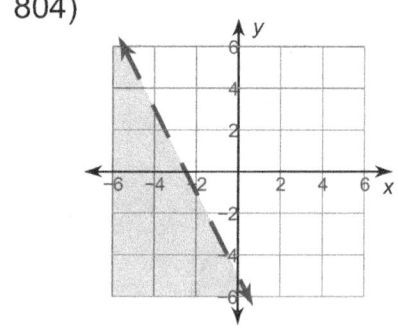

805)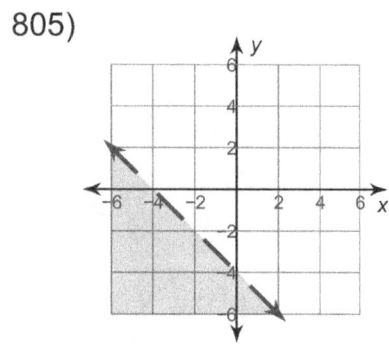

806)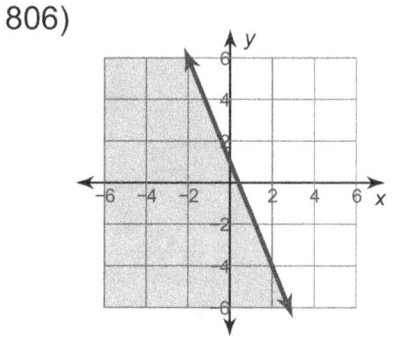

807)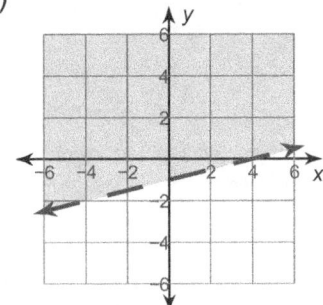

808)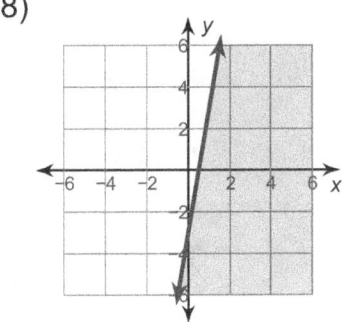

809)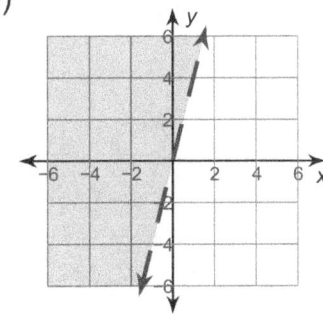

810)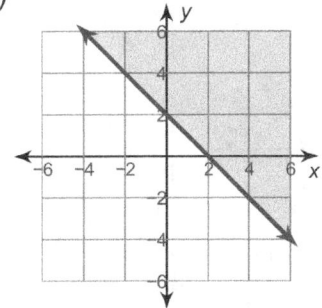

811)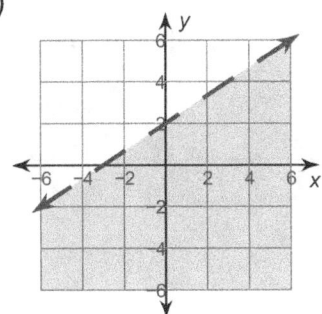

812)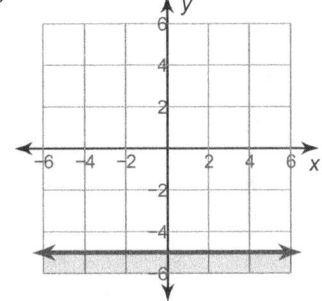

813)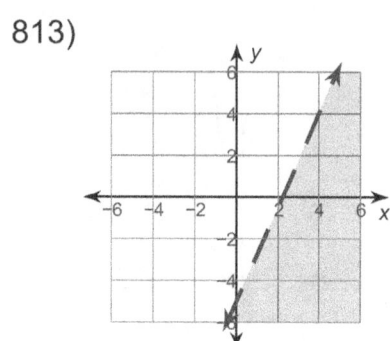

814)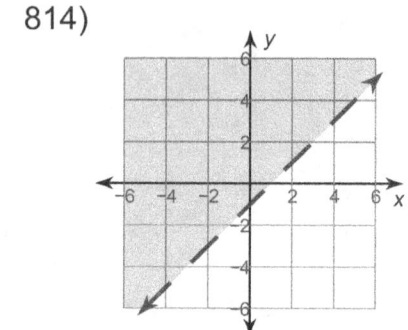

815)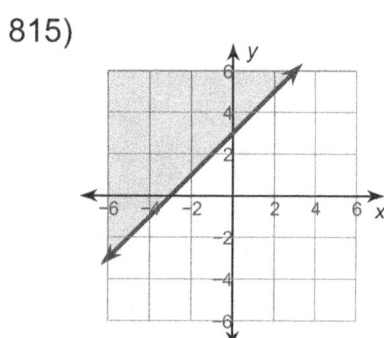

816)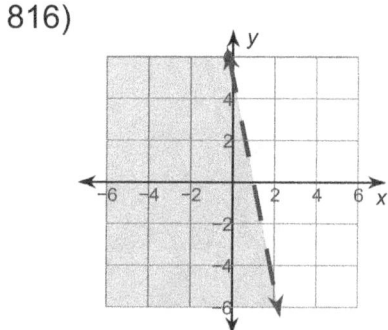

817)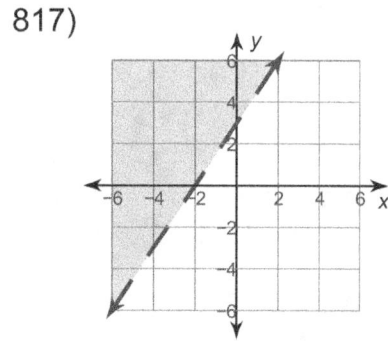

818)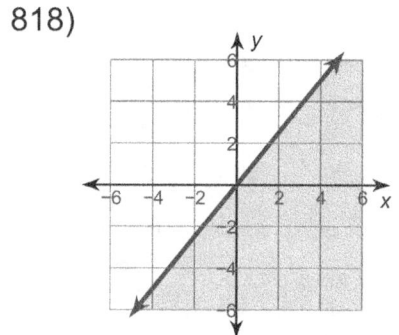

819)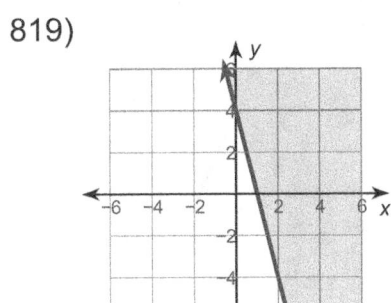

820)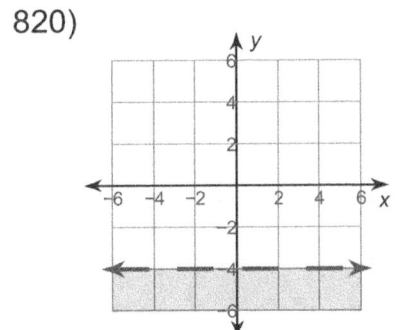

821)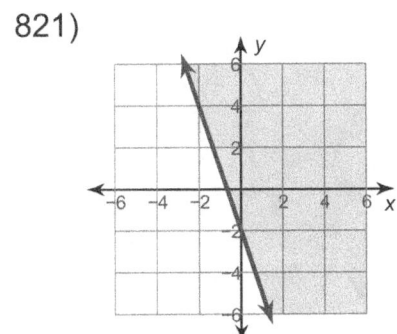

822)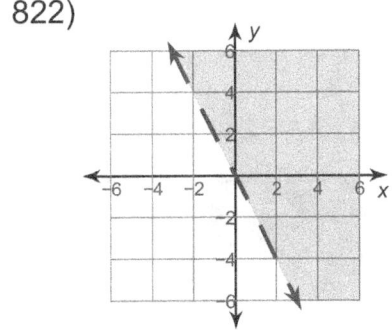

823)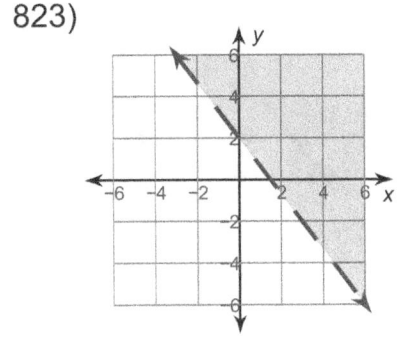

824)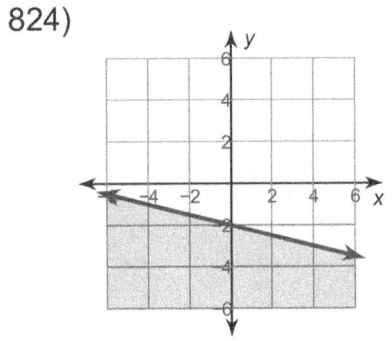

825)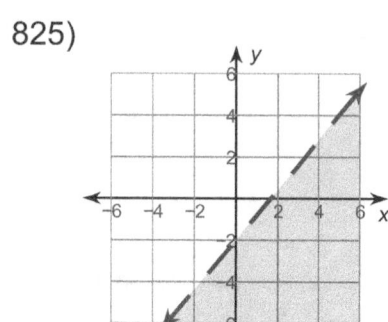

826)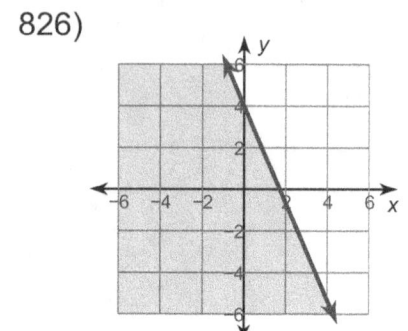

827)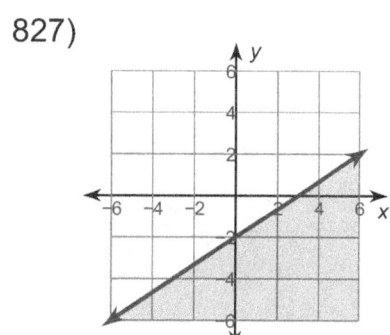

828)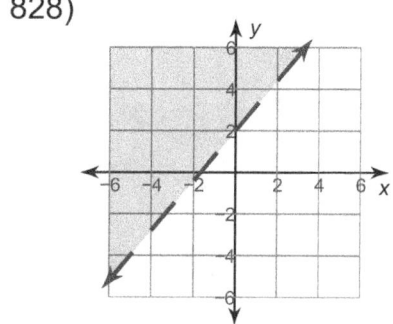

829)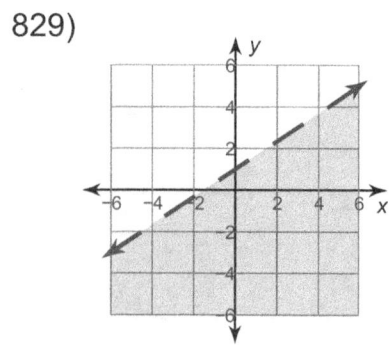

830)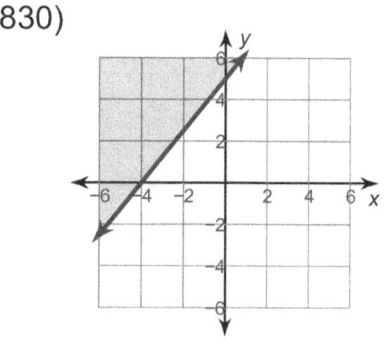

831)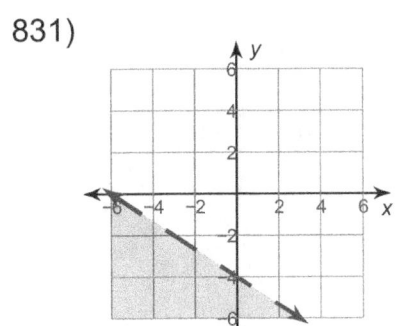

832)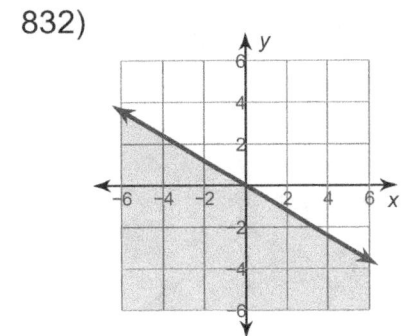

833)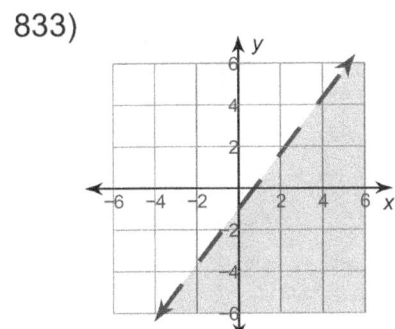

834)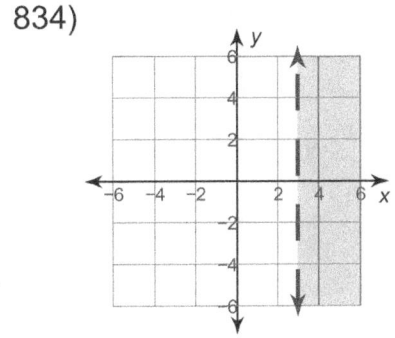

835)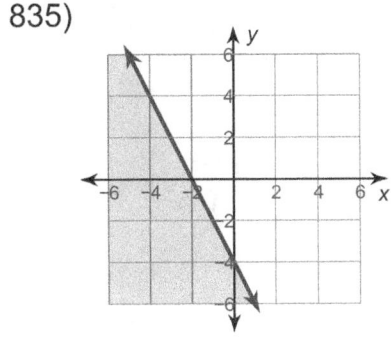

836)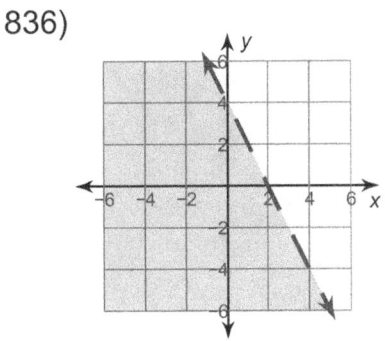

837)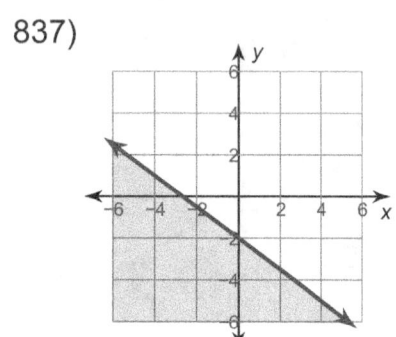

838)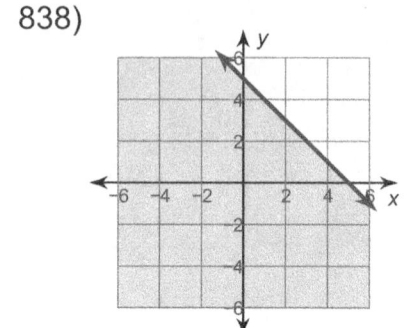

839)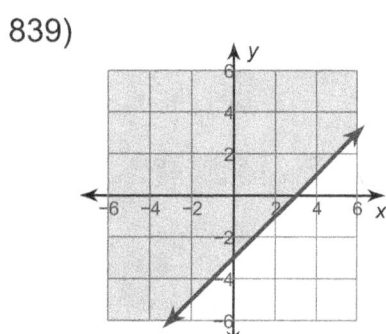

840)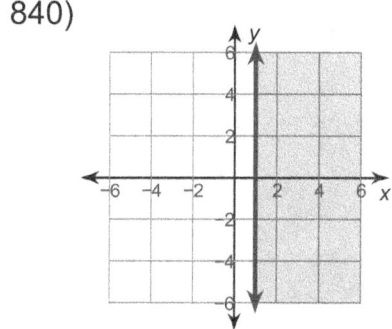

841)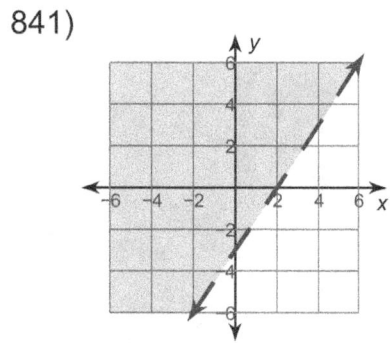

842)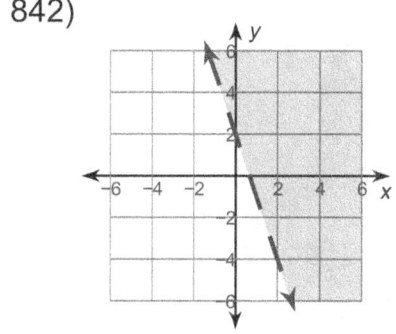

843)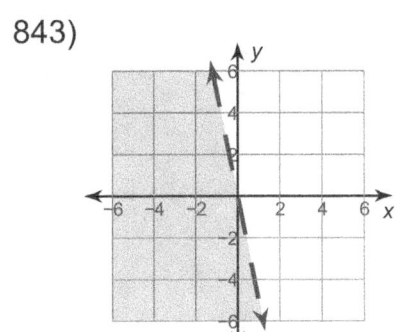

844)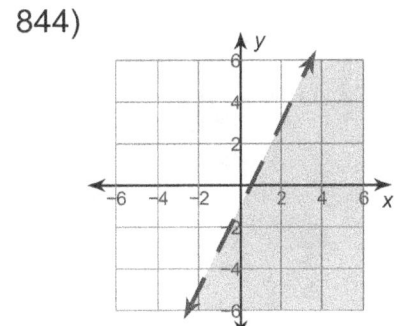

845)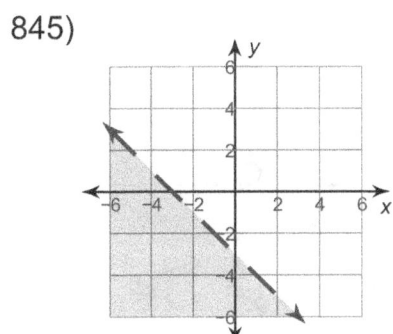

846)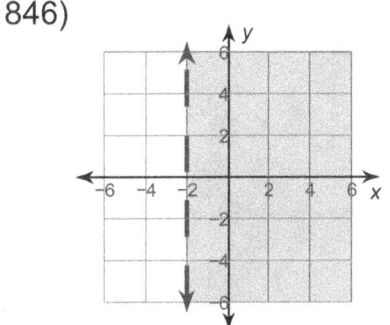

847)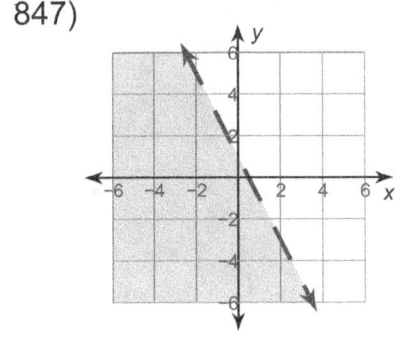

848)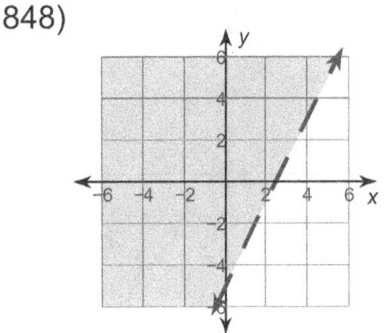

849)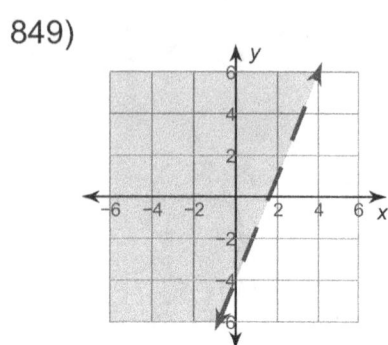

850)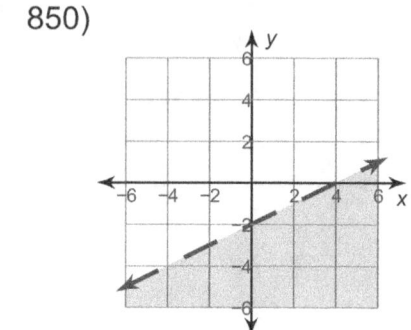

851)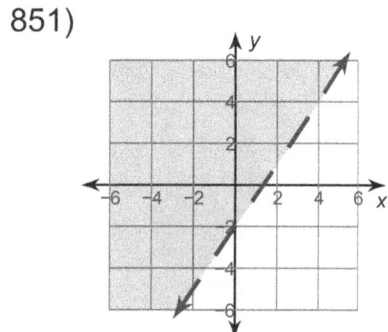

852)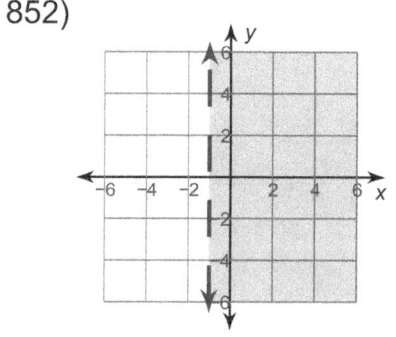

853)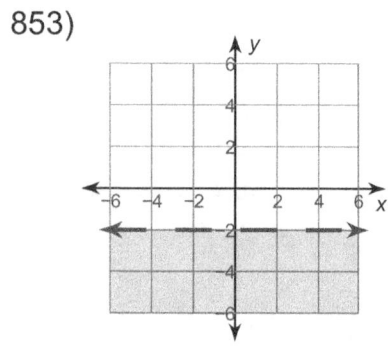

854)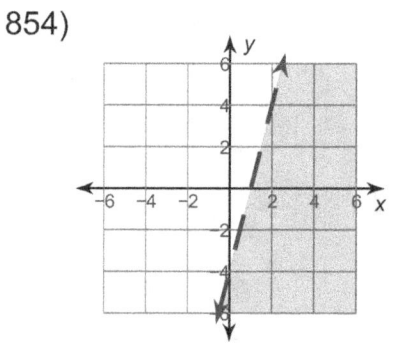

855)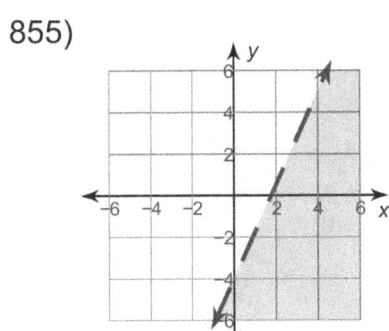

856)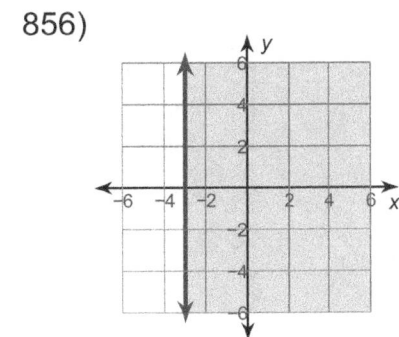

857)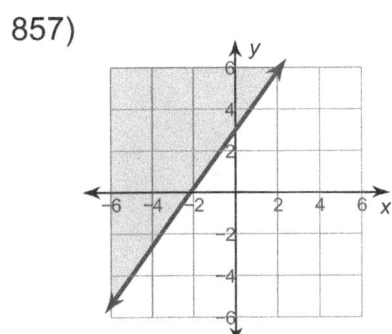

858)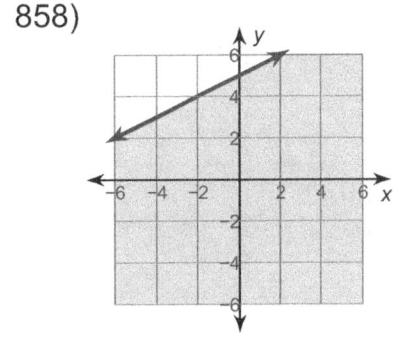

859)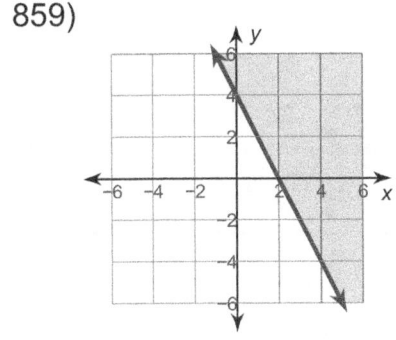

860)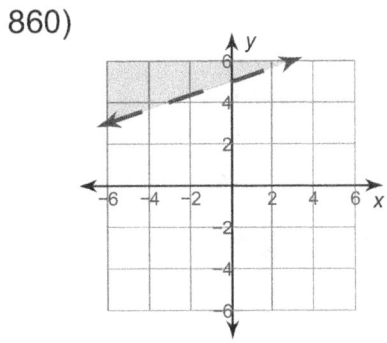

861)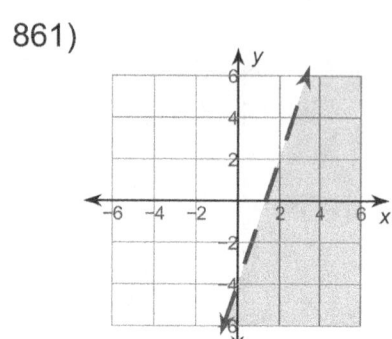

862)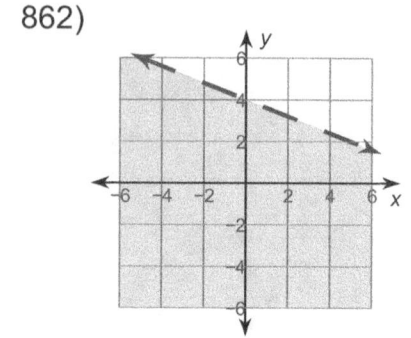

863)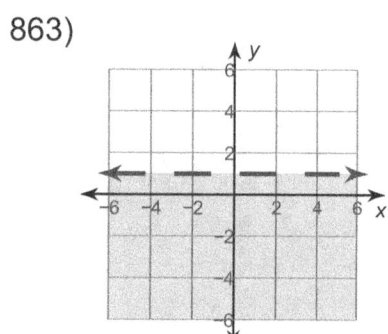

864)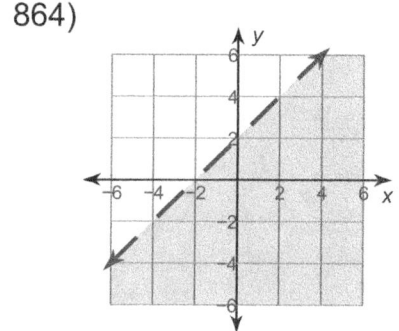

865)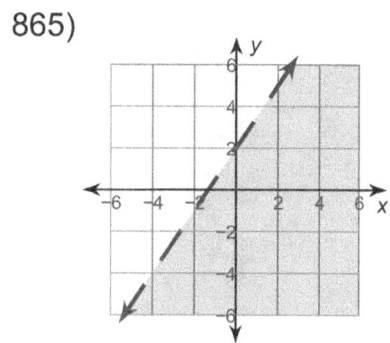

866)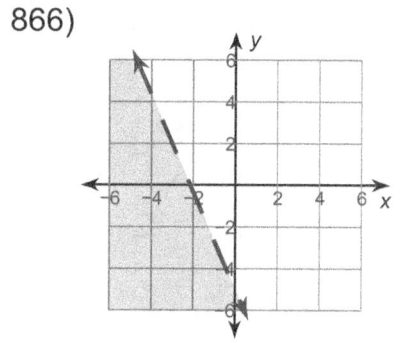

867)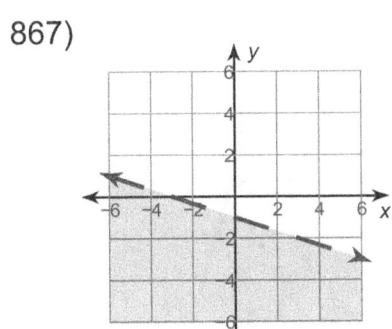

868)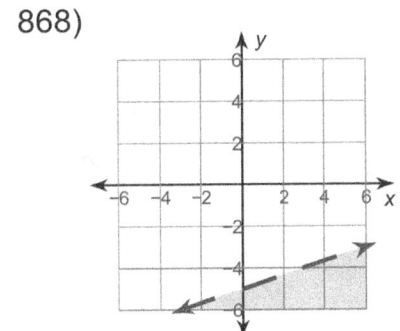

869)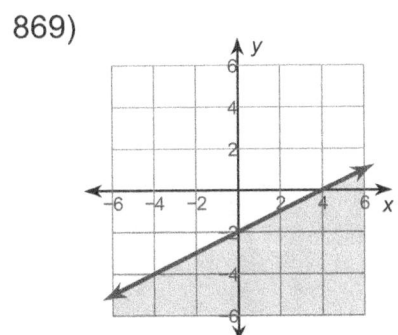

870)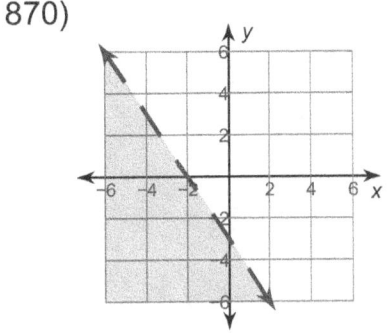

871)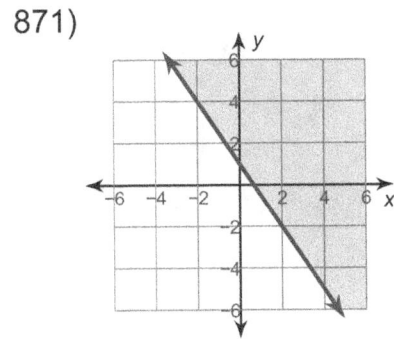

872)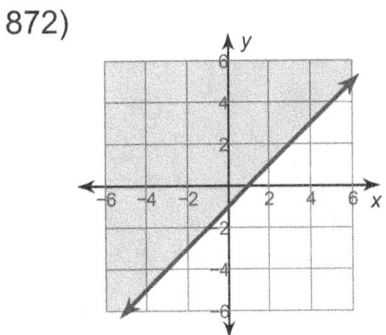

873)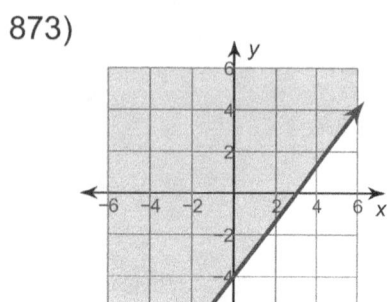

874)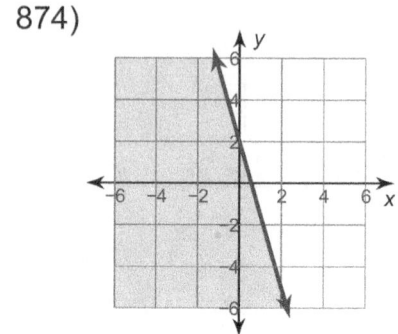

875)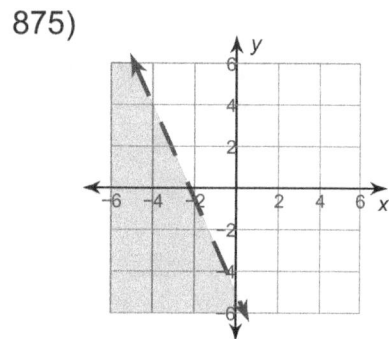

876)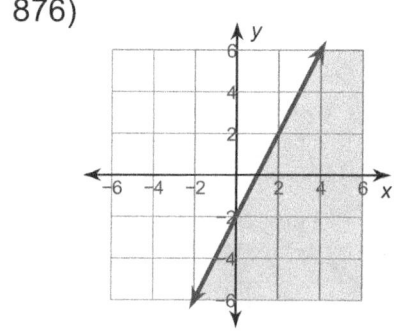

877)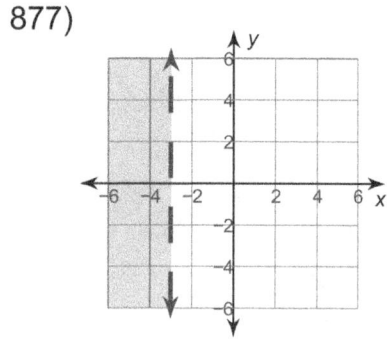

878)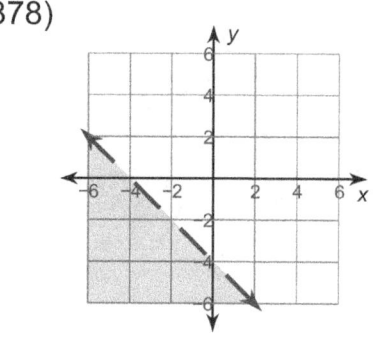

879)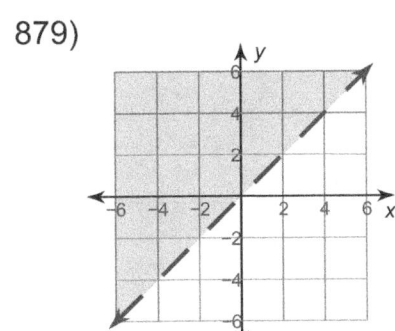

880)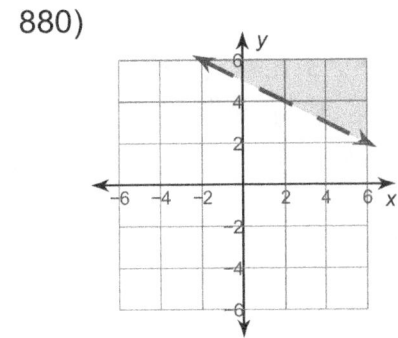

881)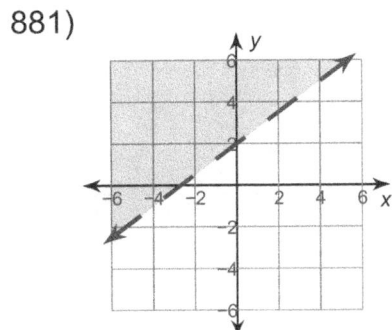

882)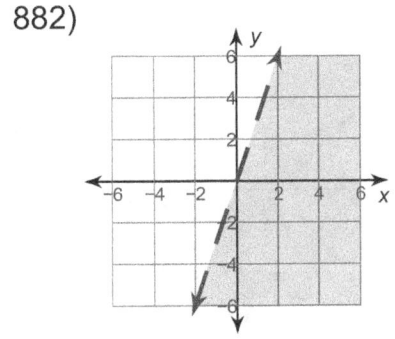

883)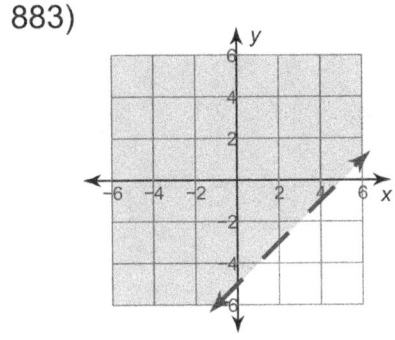

884)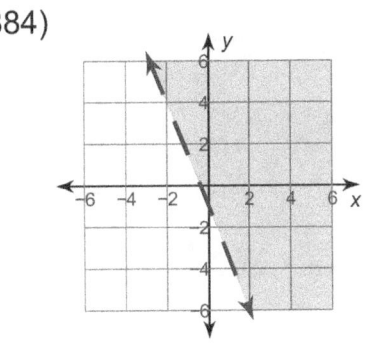

885)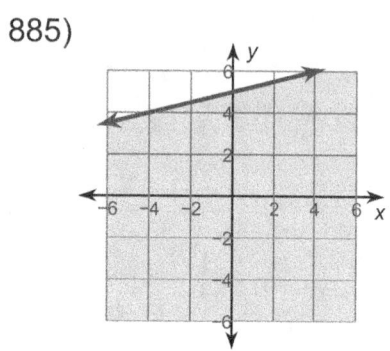

886)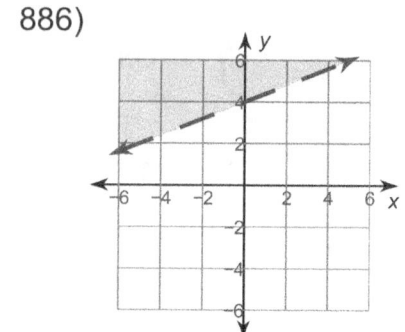

887)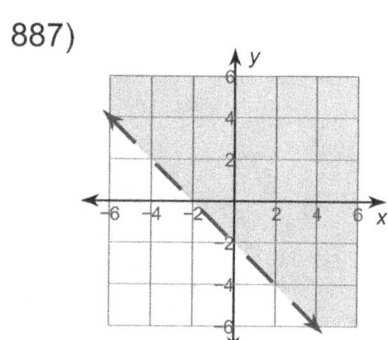

888)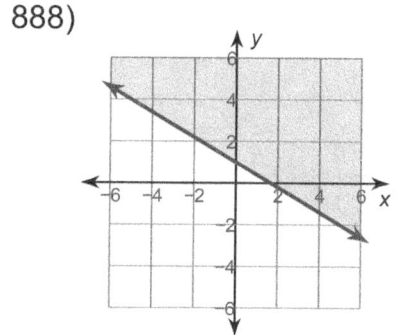

890)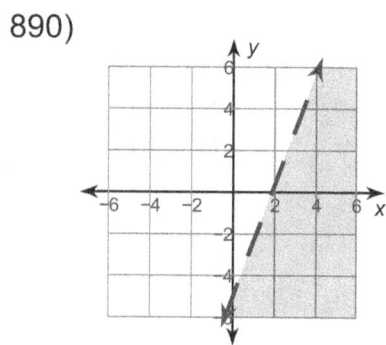

889)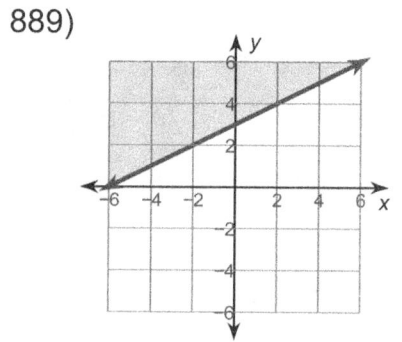

891)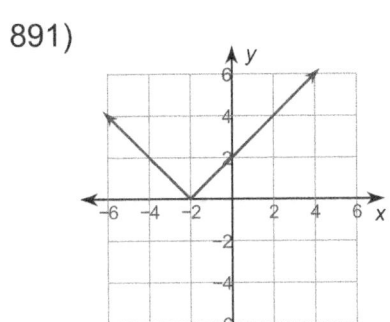

892)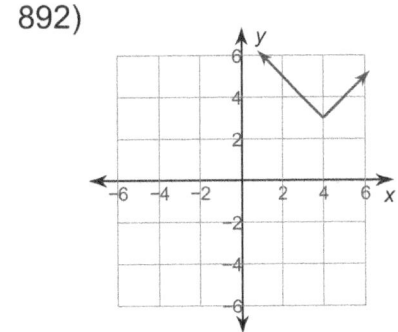

893)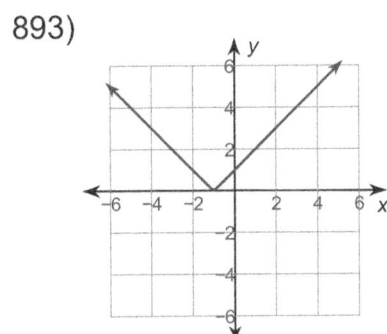

894)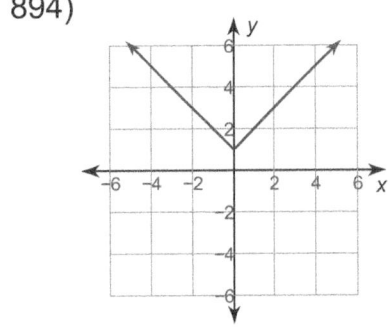

895)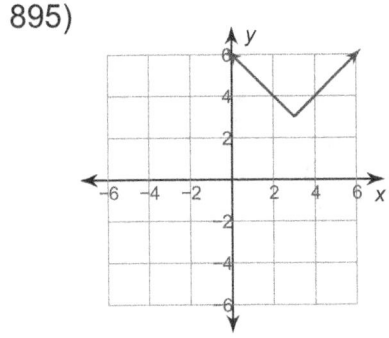

896)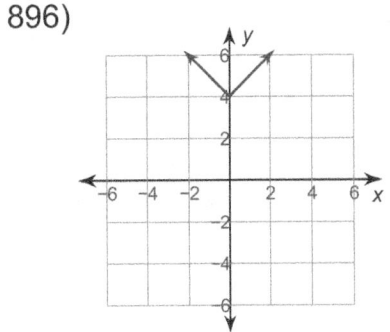

897)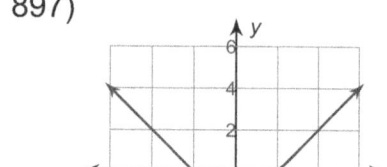

898)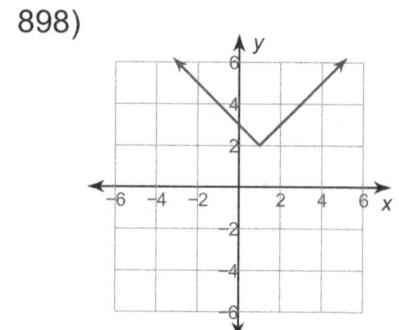

899)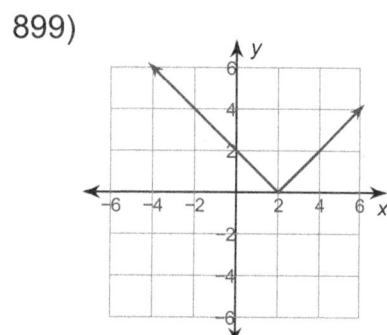

900)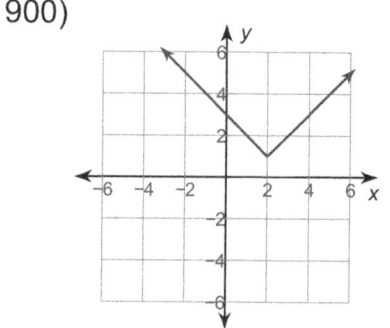

902)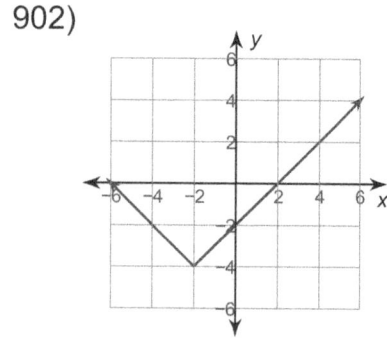

901)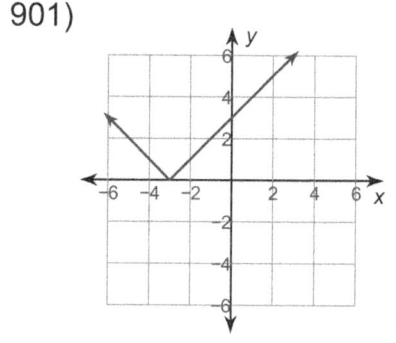

903)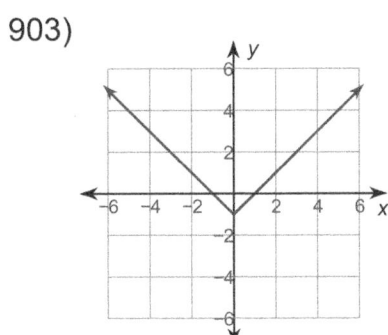

904)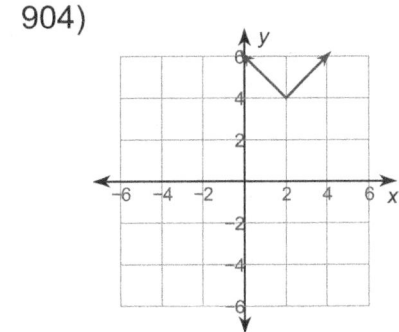

905)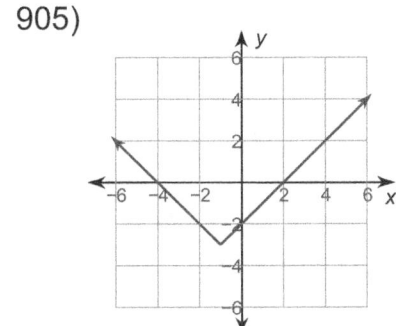

906)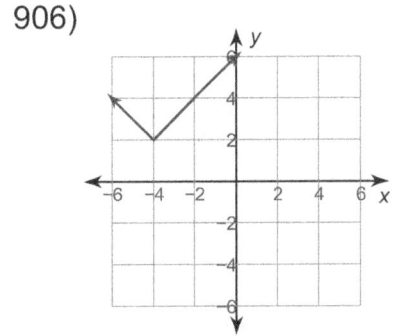

908)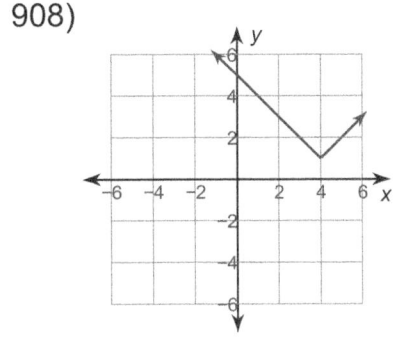

907)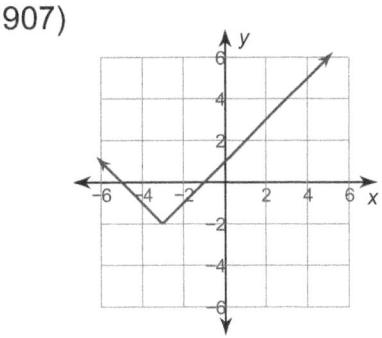

909)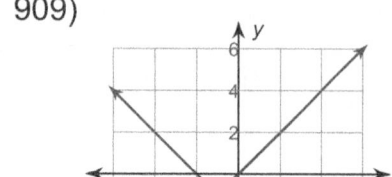

910)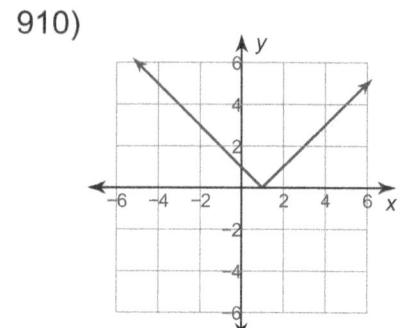

911)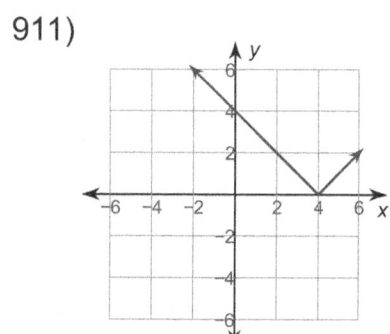

912)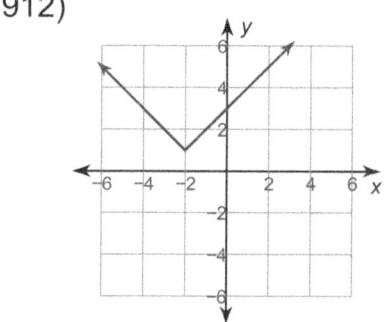

914)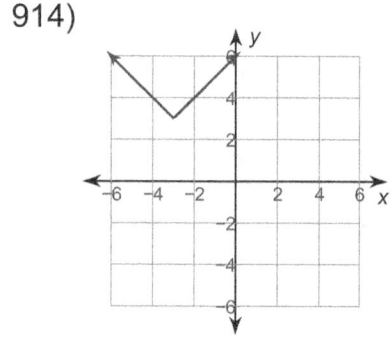

913)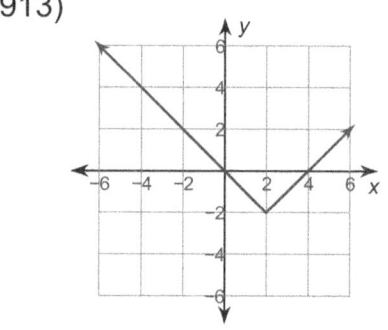

915)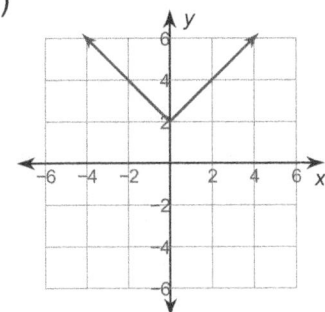

916)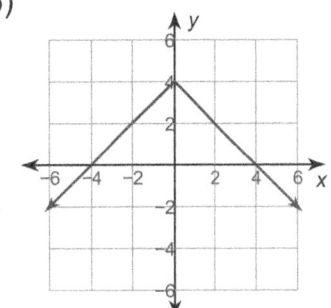

917)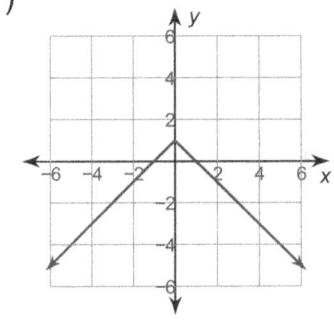

918)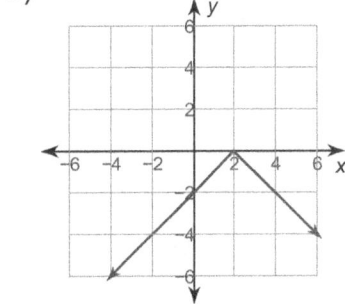

920)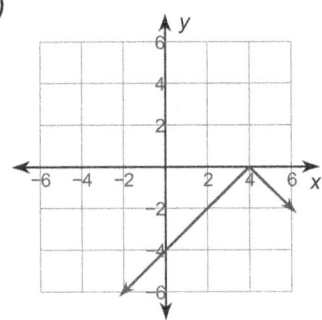

919)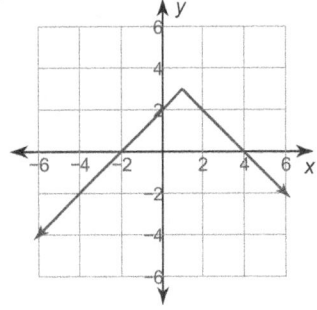

921)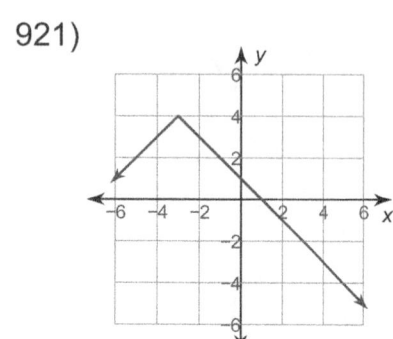

922)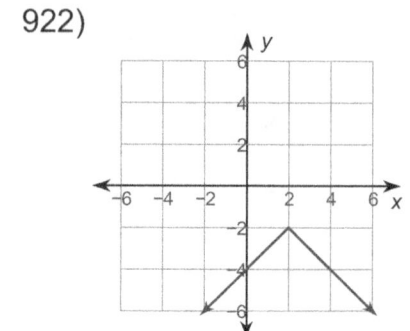

923)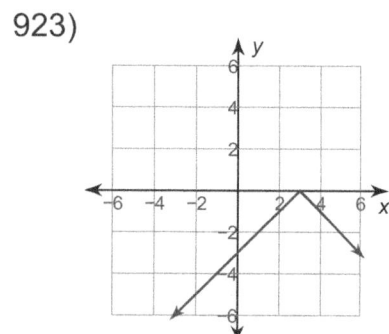

924)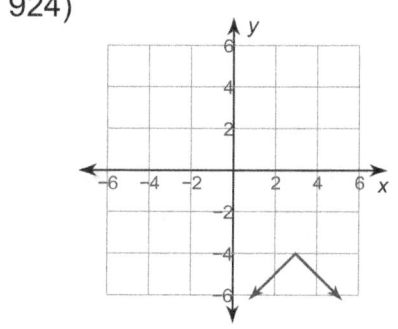

925)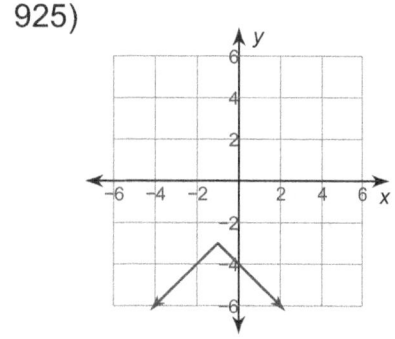

926)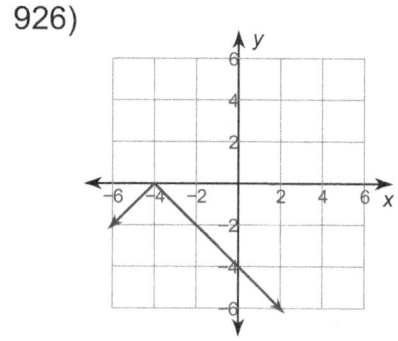

927)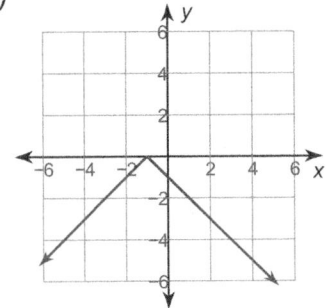

928)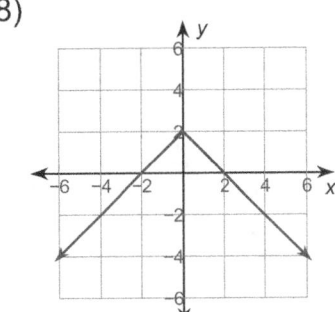

929)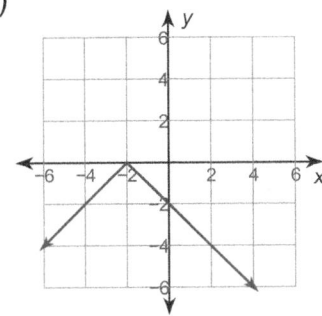

930)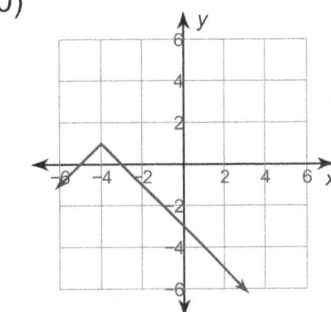

931)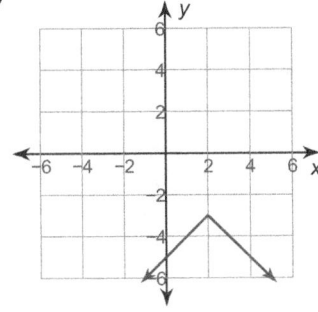

932)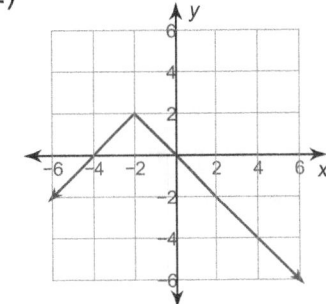

933)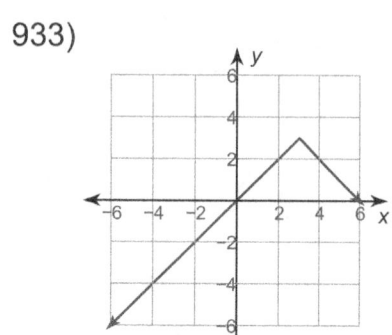

934)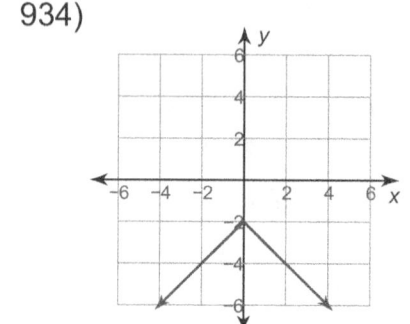

935)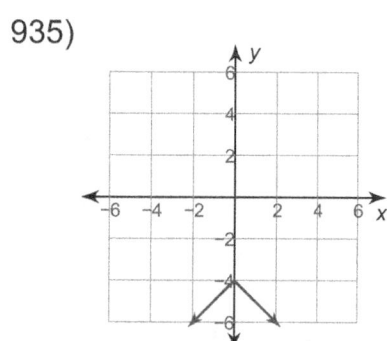

936)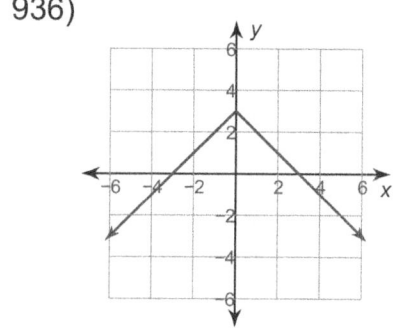

937)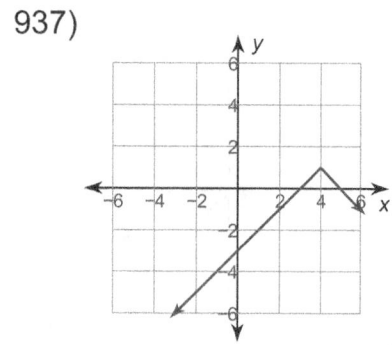

938)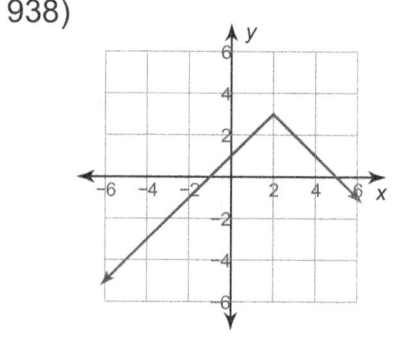

939)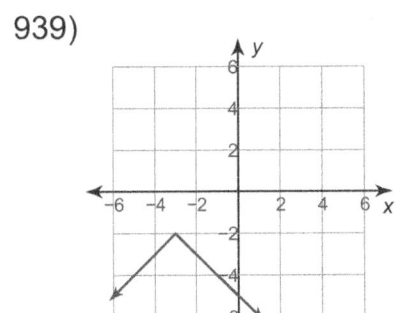

940)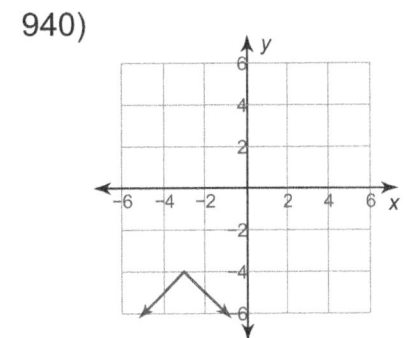

941) (–4, –3)

942) (4, 3)

943) (1, –2)

944) No solution

945) (2, –4)

946) (–1, 4)

947) (–2, 4)

948) (–2, –4)

949) (2, 2)

950) (–3, 4)

951) (–2, –2)

952) (–1, 1)

953) (2, −4) 954) (−3, 1)

955) No solution 956) No solution

957) (−3, 1) 958) (4, 4)

959) (−2, 4) 960) (1, 2)

961) No solution 962) (−4, 2)

963) (1, 3) 964) (1, −2)

965) (4, 1) 966) (1, −1)

967) No solution 968) (1, −3)

969) No solution 970) (−4 −3)

ADVANCED ALGEBRA 1

Volume 1 Answer Keys

971) (–9 –3)

972) (–10 9)

973) (–11, 18)

974) (17, –3)

975) (–10, –17)

976) (14, 2)

977) (–9, –4)

978) (14, 2)

979) (3, 12)

980) (–5, –2)

981) No solution

982) No solution

983) (14, 15)

984) (9, 12)

985) (19, 9)

986) (4, 19)

987) (9, –12)

988) (3, –16)

989) (14, 9) 990) (−14, −12)

991) (−1, 4) 992) No solution

993) (10, 14) 994) (1, 0)

995) (−2, −2) 996) No solution

997) (3, 0) 998) (4, −3)

999) (−5, 2) 1000) (10, 15)

1001) (3, 8) 1002) (15, −4)

1003) (−6, 3) 1004) (4, 3)

1005) No solution 1006) (−1, −1)

1007) (1, 5)

1008) (−7, −3)

1009) (2, 2)

1010) (9, −7)

1011) (4, 2)

1012) (2, 2)

1013) (−2, 4)

1014) (1, 1)

1015) (6, 6)

1016) (18, 17)

1017) (−3, 2)

1018) (−3, −4)

1019) (2, 2)

1020) Infinite number of solutions

1021) (1, −3)

1022) (9, 15)

1023) No solution

1024) (13, −8)

1025) (–2, 2)

1026) (5, –5)

1027) (–3, 0)

1028) (–7, 7)

1029) (2, –1)

1030) (–8, 16)

1031)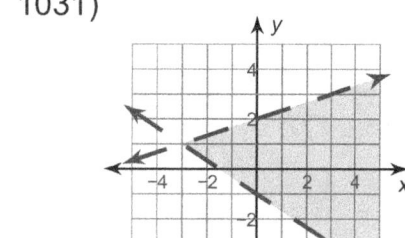

1032)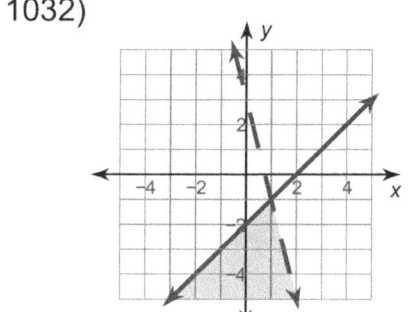

1033)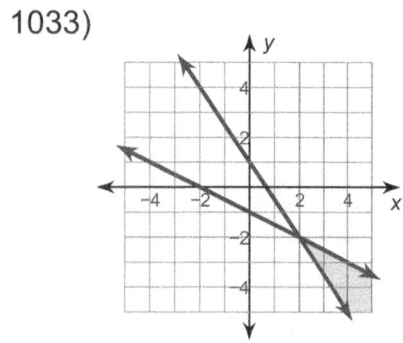

1034)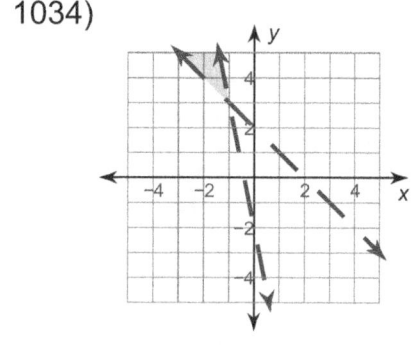

1035)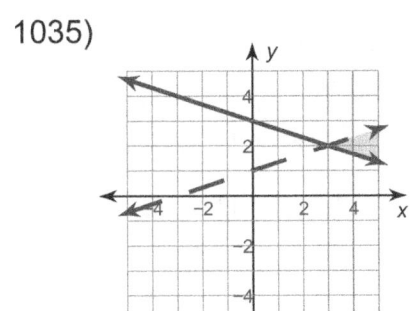

1036)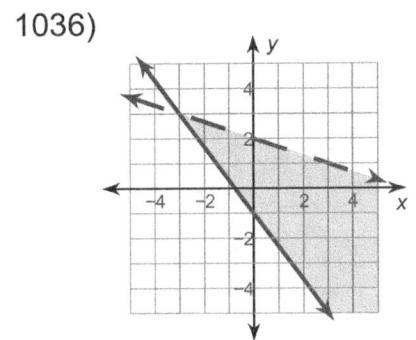

1037)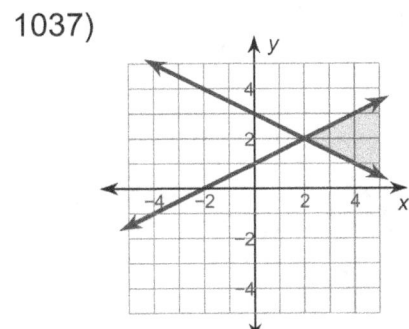

1038)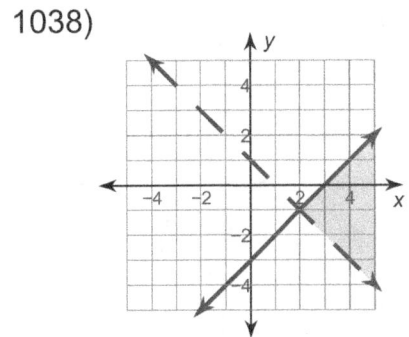

1039)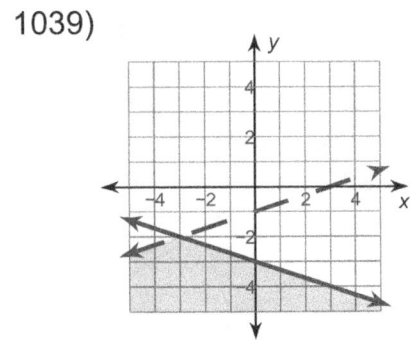

1040)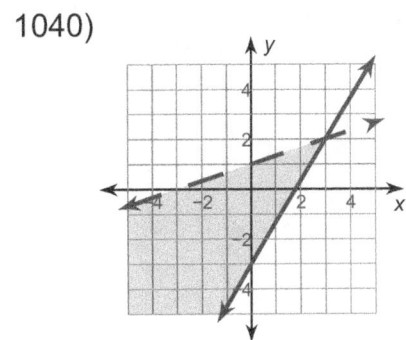

1041)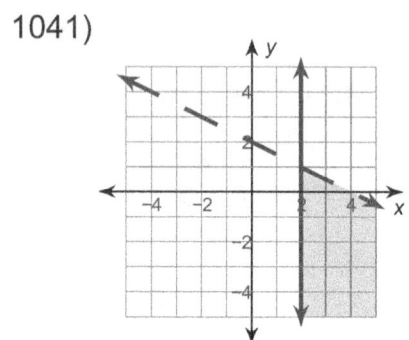

1042)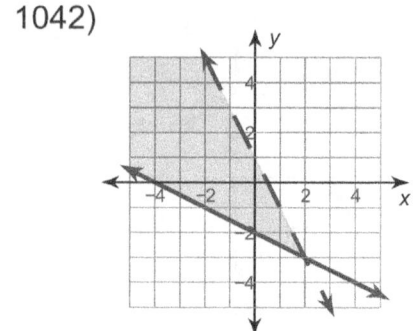

1043)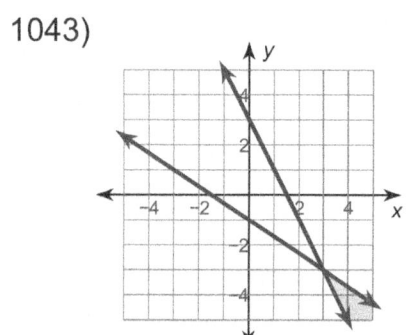

1044)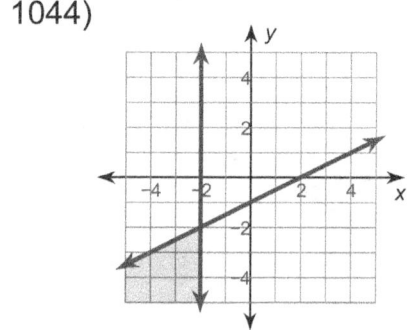

1045)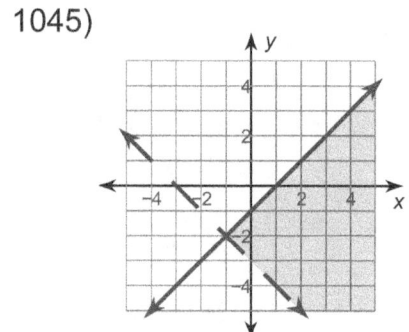

1046)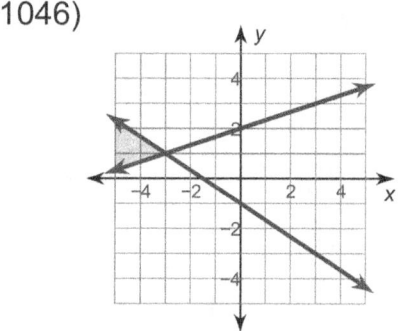

1047)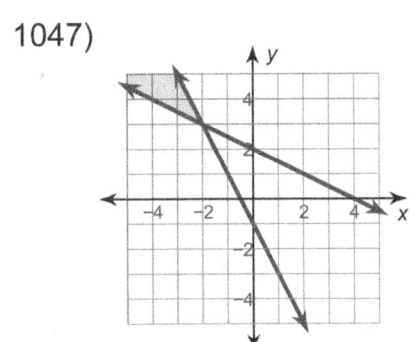

1048)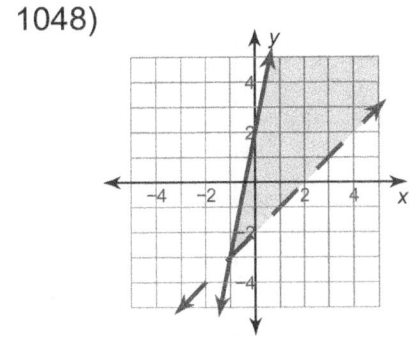

1049)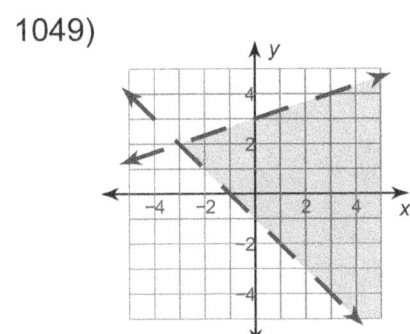

1050)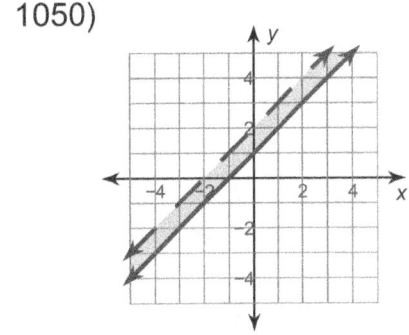

1051)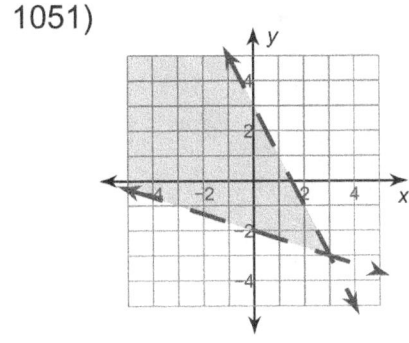

1052)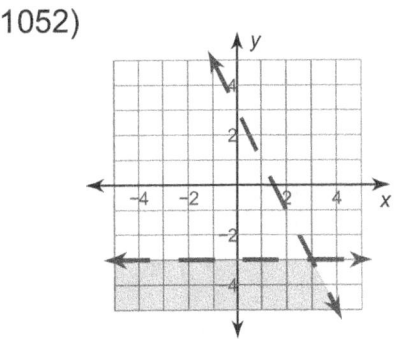

1053)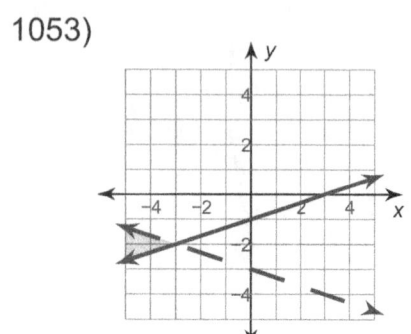

1054)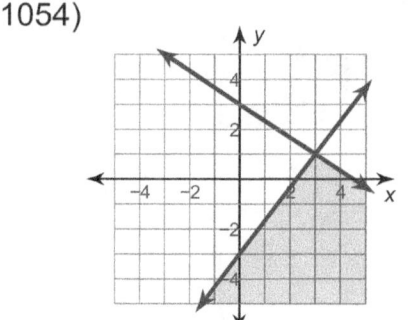

1055)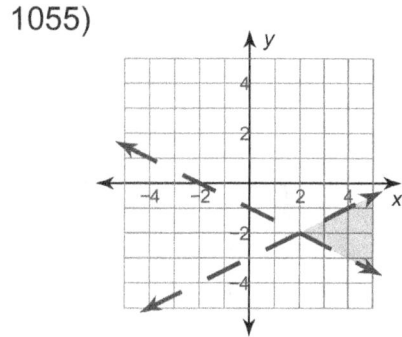

www.ingramcontent.com/pod-product-compliance
Lightning Source LLC
Chambersburg PA
CBHW081840230426
43669CB00018B/2771